ELECTRON MICROPROBE ANALYSIS AND
SCANNING ELECTRON MICROSCOPY IN GEOLOGY

This book describes electron microprobe analysis (EMPA) and scanning electron microscopy (SEM) specifically from a geological viewpoint. No prior knowledge is assumed and unnecessary technical detail is avoided, in order to keep the book easily accessible to new users of these techniques.

The principles of electron–specimen interactions and instrumentation are covered in the first part of the book. The mechanisms involved in SEM (secondary and backscattered electron) image formation are then explained, with full consideration of digital imaging techniques. The operating principles of energy- and wavelength-dispersive X-ray spectrometers are described, as well as ancillary techniques including cathodoluminescence (CL) and electron backscatter diffraction (EBSD). Procedures for qualitative and quantitative X-ray analysis (using either electron microprobe or SEM instruments) are described in detail. The production of X-ray 'maps' showing element distributions is also described, with examples. Finally the subject of specimen preparation is discussed. There is an emphasis throughout on specifically geological aspects not covered in books aimed at a more general readership.

This updated version of the first (1996) edition takes full account of recent developments and is intended for geological graduate students and postdoctoral workers, as well as those in commercial laboratories. It is also an invaluable accompaniment to courses for geological EMPA and SEM users.

D R R E E D is affiliated to the Department of Earth Sciences at the University of Cambridge. After studying physics at Southampton University, he gained a Ph.D. from the University of Cambridge in 1964 for research in using EMPA to analyse iron meteorites. He went on to be a Scientific Officer at the Natural History Museum, London from 1965 until 1970 before his appointment as Senior Research Fellow at the Australian National University, Canberra in 1970, where he implemented a new system for quantitative ED analysis. From

1974 until his retirement in 2002, Dr Reed was at the Department of Earth Sciences, University of Cambridge with research interests including ion and electron microprobe analysis and developing simulation software. In 1981 he was awarded the Microbeam Analysis Society Presidential Award for his outstanding scientific contribution to the theory and practice of microbeam analysis, followed in 1984 by honorary life membership. He has written, and contributed to, several books on the subject, including *Electron Microprobe Analysis* (Cambridge University Press, first edn 1975, second edn 1993).

ELECTRON MICROPROBE ANALYSIS AND SCANNING ELECTRON MICROSCOPY IN GEOLOGY

S. J. B. REED

University of Cambridge

CAMBRIDGE
UNIVERSITY PRESS

CAMBRIDGE UNIVERSITY PRESS
Cambridge, New York, Melbourne, Madrid, Cape Town, Singapore,
São Paulo, Delhi, Dubai, Tokyo

Cambridge University Press
The Edinburgh Building, Cambridge CB2 8RU, UK

Published in the United States of America by Cambridge University Press, New York

www.cambridge.org
Information on this title: www.cambridge.org/9780521142304

First published 2005
This digitally printed version 2010

A catalogue record for this publication is available from the British Library

ISBN 978-0-521-84875-6 Hardback
ISBN 978-0-521-14230-4 Paperback

Additional resources for this publication at www.cambridge.org/9780521142304

Contents

The colour plates are situated between pages 96 and 97.*
*These plates are available for download in colour from
www.cambridge.org/9780521142304

Preface

The favourable reception given to the first (1996) edition of this book suggests that the joint treatment of electron microprobe analysis (EMPA) and scanning electron microscopy (SEM) with a specifically geological slant has been found to serve a useful purpose. It was therefore decided to proceed with this second, revised and updated, edition. The inclusion of both EMPA and SEM can be justified on the grounds that the instruments share much in common and their functions overlap: SEMs fitted with X-ray spectrometers are often used in analytical mode, while EMP instruments, though designed primarily for analysis, also have imaging functions similar to those of the SEM.

The capabilities of the computers used both for instrument control and for data processing have increased greatly since the first edition. Whilst this allows more sophisticated software functions, it does not diminish the need to understand both the operating principles of the instruments and the factors controlling the results, the explanation of which is the main purpose of this book. Digital rather than analogue imaging is now the norm, with concomitant advantages provided by image processing and image analysis techniques. The increasing use of 'false' colour images in various forms is reflected in an expanded colour section in this edition. Significant instrumental developments include the increasing adoption of field emission electron sources, which are especially beneficial for high-resolution SEM applications. Also, variable-pressure or environmental SEMs are more commonly used. In addition, interest in ancillary techniques such as cathodoluminescence and electron backscatter diffraction has grown.

As before, no prior knowledge is expected of the reader and technical detail is limited to that needed for a sound understanding of operating principles and interpretation of results. It is hoped that the book will be particularly useful to postgraduate students and postdoctoral researchers in university geology departments, where it may serve as an accompaniment to courses for SEM and EMPA users.

Inevitably a book reflects the bias of the author and for this I ask the reader's indulgence, as well as for any errors or omissions.

Acknowledgments

I am greatly indebted to the following for providing illustrative material: J. Barreau (Fig. 4.21), N. J. Butterfield (Fig. 4.10) and J. A. D. Dickson (Figs. 4.7 and 4.19), Department of Earth Sciences, University of Cambridge; N. Cayzer (Figs. 4.8, 4.23 and cover), S. Haszeldine (Fig. 4.11) and N. Kelly (Fig. 4.33), Department of Geology and Geophysics, University of Edinburgh; T. J. Fagan (Plate 7), School of Ocean Science and Technology, University of Hawai'i at Manoa; B. J. Griffin (Fig. 4.34), Centre for Microscopy and Microanalysis, University of Western Australia; M. Jercinovic and M. Williams (Plates 5 and 6), Department of Geosciences, University of Massachusetts; M. Lee (Fig. 4.32), Division of Earth Sciences, University of Glasgow; G. E. Lloyd (Plate 3), Department of Earth Sciences, University of Leeds; E. W. Macdonald (Fig. 4.9), Department of Earth Sciences, Dalhousie University; A. Markowitz and K. L. Milliken (Plate 4(a)) and R. M. Reed (Plate 4(b)), Department of Geological Sciences, University of Texas at Austin; F. S. Spear and C. G. Daniel (Plate 8), Department of Earth and Environmental Sciences, Rensselaar Polytechnic Institute; P. D. Taylor (Fig. 4.18), Department of Palaeontology, Natural History Museum, London; and P. Trimby (Fig. 4.31), HKL Technology, Hobro, Denmark.

Copyright permission was kindly granted by the following: Mineralogical Society of America (Plate 8); Paleontological Society (Fig. 4.9); *Journal of Sedimentary Research* (Plate 4(a)); *Meteoritics and Planetary Sciences* (Plate 7); and *Microscopy and Analysis* (Fig. 4.32).

On a personal note, I would like to record my indebtedness to Jim Long (1926–2003), who played a pivotal role in the development of EMPA in Britain, and whose knowledge and wisdom are greatly missed.

1

Introduction

1.1 Electron microprobe analysis

Electron microprobe analysis (EMPA) is a technique for chemically analysing small selected areas of solid samples, in which X-rays are excited by a focussed electron beam. (The term 'electron probe microanalysis', or EPMA, is synonymous.) The X-ray spectrum contains lines characteristic of the elements present; hence a qualitative analysis is easy to obtain by identifying the lines from their wavelengths (or photon energies). By comparing their intensities with those emitted from standard samples (pure elements or compounds of known composition) it is also possible to determine the concentrations of the elements quantitatively. Accuracy approaching $\pm 1\%$ (relative) is obtainable and detection limits down to tens of parts per million (by weight) can be attained. Under normal conditions, spatial resolution is limited to about 1 μm by the spreading of the beam within the sample. The spatial distributions of specific elements can be recorded in the form of line profiles or two-dimensional 'maps', which are commonly displayed using a 'false' colour scale to represent elemental concentrations.

1.2 Scanning electron microscopy

The scanning electron microscope (SEM) is a close relative of the electron microprobe (EMP) but is designed primarily for imaging rather than analysis. Images are produced by scanning the beam while displaying the signal from an electron detector on a TV screen or computer monitor. By choosing the appropriate detection mode, either topographic or compositional contrast can be obtained. ('Composition' here refers to mean atomic number: individual elements cannot be distinguished.) Spatial resolution better than 10 nm in topographic mode and 100 nm in compositional mode can be achieved, though

1

in many applications the large depth of field in SEM images (typically at least 100 times greater than for a comparable optical microscope) is more relevant than high resolution. An important factor in the success of the SEM is that images of three-dimensional objects are usually amenable to immediate intuitive interpretation by the observer. The range of applications of SEM can be extended by adding other types of detector, e.g. for light emission caused by electron bombardment, or cathodoluminescence (CL).

1.2.1 Use of SEM for analysis

Scanning electron microscopes commonly have an X-ray spectrometer attached, enabling the characteristic X-rays of a selected element to be used to produce an image. Also, with a stationary beam, point analyses can be obtained, as in EMPA. (The spatial resolution with respect to analysis is, however, still limited to about 1 μm by beam spreading, despite the higher resolution obtainable in scanning images.) Since EMP instruments have electron imaging facilities, used primarily for locating points for analysis, the functions of the two instruments overlap considerably. The SEM is optimised for imaging, with analysis as an extra, whereas in the EMP the priorities are reversed and various additional features that facilitate analysis are incorporated.

1.3 Geological applications of SEM and EMPA

The advantages of the SEM as an imaging instrument (high spatial resolution, large depth of field, and simple specimen preparation) make it an invaluable tool in the following branches of geology.

> *Palaeontology.* The SEM is ideally suited to the study of fossil morphology, especially that of micro-fossils.
> *Sedimentology.* Three-dimensional images of individual sediment grains and intergrowths can be obtained; data on fabric and porosity can also be generated.
> *Mineralogy.* The SEM is very effective for studying crystal morphology on a microscale.
> *Petrology.* The ability to produce images of polished sections showing differences in mean atomic number is very useful both in sedimentary and in igneous petrology.

The reasons for the widespread application of EMPA to geology (whether carried out in a 'true' EMP instrument or SEM with X-ray spectrometer fitted), especially in the fields of mineralogy and petrology, can be summarised as follows.

(1) Specimen preparation is straightforward and entails the use of existing techniques of section-making and polishing with only minor modifications.
(2) The technique is non-destructive, unlike most other analytical techniques.
(3) Quantitative elemental analysis with accuracy in the region of $\pm 1\%$ (for major elements) can be obtained.
(4) All elements above atomic number 3 can be determined (with somewhat varying accuracy and sensitivity).
(5) Detection limits are low enough to enable minor and trace elements to be determined in many cases.
(6) The time per analysis is reasonably short (usually between 1 and 5 min).
(7) Spatial resolution of the order of 1 µm enables most features of interest to be resolved.
(8) Individual mineral grains can be analysed *in situ*, with their textural relationships undisturbed.
(9) A high specimen throughput rate is possible, the time required for changing specimens being quite short.

These characteristics have proved useful in the following subject areas.

Descriptive petrology. The EMPA technique is commonly used for the petrological description and classification of rocks and has an importance comparable to that of the polarising microscope.

Mineral identification. As an adjunct to polarised-light microscopy and X-ray diffraction, EMPA provides compositional information that assists in mineral identification.

Experimental petrology. For experimental studies on phase relationships and elemental partitioning between coexisting phases, the spatial resolution of the electron microprobe is especially useful, given the typically small grain size.

Geothermobarometry. The EMPA technique is ideally suited to the determination of the composition of coexisting phases in rocks, from which temperatures and pressures of formation can be derived.

Age determination. Th–U–Pb dating of minerals containing insignificant amounts of non-radiogenic Pb (such as monazite) is possible by EMPA, with higher spatial resolution than can be obtained with isotopic methods, though lower accuracy.

Zoning. The high spatial resolution of the technique enables zoning within mineral grains to be studied in detail.

Diffusion studies. Experimental diffusion profiles in geologically relevant systems can be determined with the electron microprobe, its high spatial resolution being crucial in this field.

Modal analysis. Volume fractions of minerals and other data can be obtained by automated modal analysis, mineral identification being based on X-ray and sometimes backscattered-electron signals.

Rare-phase location. Grains of rare phases can be located by automated search procedures, using the X-ray signal for one or more diagnostic elements.

1.4 Related techniques

Though EMPA has many useful attributes, as described in the previous section, other, complementary analytical methods offer advantages in one respect or another. These are outlined briefly in the following sections.

1.4.1 Analytical electron microscopy

With specimens less than 100 nm thick and an electron energy of at least 100 keV, much better spatial resolution (down to 10 nm) can be obtained, owing to the relatively small amount of lateral scattering which occurs as the electrons pass through the specimen. The reduction in X-ray intensity can be compensated to a large extent by using an X-ray detector of high collection efficiency and a high-intensity electron source. This type of analysis can be carried out in an 'analytical electron microscope' (AEM), in which conventional electron transmission imaging and diffraction capabilities are combined with X-ray detection. Analysis with high spatial resolution is also possible with a scanning transmission electron microscope (STEM) fitted with an X-ray spectrometer.

   Another analytical technique that is also available in AEM and STEM instruments is electron energy-loss spectrometry ('EELS'), which utilises steps in the energy spectrum of transmitted electrons caused by energy losses associated with inner-shell ionisation. Electron spectrometers with parallel collection make this a very sensitive technique.

   For further information about AEM, see Joy, Romig and Goldstein (1986) and Champness (1995).

1.4.2 Proton-induced X-ray emission

Characteristic X-rays can be excited by bombardment with protons, giving rise to the technique known as 'PIXE' (proton-induced X-ray emission). The principal advantage of PIXE is that the X-ray background is much lower than in EMPA (a consequence of the higher mass of the proton compared with the electron), making small peaks easier to detect. Detection limits are thus typically an order of magnitude lower (in the ppm range). On the other hand, high-energy protons are more difficult to focus in order to obtain high spatial resolution (a beam diameter of 1 μm is attainable, but only with low current), and they penetrate much further in solid materials. Protons of energy 1–4 MeV, which give efficient X-ray excitation, can penetrate the full 30-μm thickness of a petrological thin section and it follows that the spatial resolution with respect to *depth* is relatively poor. The equipment is quite costly and not very widely available, so geological applications have been fairly limited.

For more details about PIXE, see Fraser (1995), Halden, Campbell and Teesdale (1995) and Cabri and Campbell (1998).

1.4.3 X-ray fluorescence analysis

An alternative way of exciting characteristic X-rays is to bombard the specimen with X-rays of higher energy, this technique being known as X-ray fluorescence (XRF) analysis. It has been a standard method of elemental analysis in geology for a long time and offers good accuracy for major elements and detection limits in the region of 1 ppm. In its usual form it is a bulk method, requiring a significant amount of sample for analysis, and is therefore used principally for analysing whole rocks or separated minerals. An electron microprobe or SEM can, however, be converted to make it capable of XRF analysis with a spatial resolution of about 100 µm, by using the beam to excite X-rays in a target close to the specimen, in which fluorescent X-rays are excited.

The advent of synchrotron X-ray sources has revolutionised the possibilities of XRF analysis. The extremely high X-ray intensities available from these sources enable intense beams down to 1 µm in diameter to be produced for exciting the specimen, giving a microprobe technique in which high spatial resolution and low detection limits are combined. The accessibility of this technique is restricted by the limited number of synchrotrons in existence. For more information, see Smith and Rivers (1995). Relatively compact XRF analysers with high spatial resolution, using a low-power X-ray tube with a focussing device, are also available.

1.4.4 Auger analysis

The process known as the 'Auger effect', whereby an atom excited by electron bombardment may dissipate its energy by ejecting an electron rather than by characteristic X-ray emission, gives rise to an alternative method of analysis, which exploits the fact that the electron spectrum contains lines that have energies related to the atomic energy levels and are therefore characteristic of the element. Auger analysis is most effective for elements of atomic number below 10, for which X-ray analysis is least sensitive. Also, it differs in being a surface analysis technique, only electrons originating from depths of the order of 10 nm being detected. The scanning Auger microscope (SAM) is a close relative of the SEM, but is orientated towards the requirements of the electron energy analyser, and the instrument has to be operated at ultra-high vacuum.

This technique and its applications in geology are discussed in Section 4.8.6.

1.4.5 Ion microprobe analysis

Ion microprobe analysis is complementary to EMPA in its capabilities but has a completely different physical basis. This technique is a form of secondary-ion mass spectrometry (SIMS), whereby the specimen is bombarded with 'primary' ions, and ionised atoms from the sample ('secondary' ions) are collected and passed through a mass spectrometer. In the ion microprobe the primary ions are focussed to form a beam typically a few micrometres in diameter. For elemental analysis the advantages of the technique are low detection limits (much lower for some elements than others) and the ability to detect light elements (including H and Li) with high sensitivity. For many elements the ion microprobe lowers detection limits by several orders of magnitude compared with EMPA. Its value is enhanced by the possibility of obtaining isotopic data on small areas.

For further details, see Hinton (1995) and McMahon and Cabri (1998).

1.4.6 Laser microprobe methods

Another means by which micro-volumes of sample can be removed for analysis is bombardment with a focussed laser beam. This can be used in various ways. For example, in laser-induced mass spectrometry (LIMS) ions are generated directly by the laser beam and subjected to mass spectrometry. An alternative approach is to use less intense bombardment to generate local heating and release gases such as Ar, which can be transferred to a mass spectrometer, enabling isotope data to be obtained from selected areas. Yet another technique is to remove (or 'ablate') material and transfer this to an 'inductively coupled plasma' (ICP) source, from which either ions for mass spectrometry or light for optical emission spectrometry can be produced. The spatial resolution of these techniques is typically a few micrometres.

For further details of these techniques, see Perkins and Pearce (1995).

2

Electron–specimen interactions

2.1 Introduction

Electrons impinging on solid materials are slowed down principally through 'inelastic' interactions with outer atomic electrons, while 'elastic' deflections by atomic nuclei determine their spatial distribution. Some leave the target again, having been deflected through an angle of more than 90°. Both these 'backscattered' electrons and 'secondary' electrons dislodged from the surface of the sample are used for image formation. In addition, interactions between bombarding electrons and atomic nuclei give rise to the emission of X-ray photons with any energy up to E_0, the energy of the incident electrons, resulting in a 'continuous X-ray spectrum' (or 'continuum'). 'Characteristic' X-rays (used for chemical analysis) are produced by electron transitions between inner atomic energy levels, following the creation of a vacancy by the ejection of an inner-shell electron.

2.2 Inelastic scattering

In the SEM or EMP the electrons with which the specimen is bombarded have an energy typically in the range 5–30 keV[*], which is dissipated in interactions of various types with bound electrons and the lattice, known collectively as 'inelastic scattering'. Individual energy losses are mostly small; hence it is a reasonable approximation to assume that the electron decelerates smoothly as a function of distance travelled. The rate of energy loss is dependent on the property of the target material known as the 'stopping power', defined as $-\mathrm{d}E / \mathrm{d}(\rho s)$, where ρ is the density of the target and s the distance travelled.

---

[*] Electron and X-ray energies are expressed in electron volts (eV), 1 eV being the energy corresponding to a change of one volt in the potential of an electron.

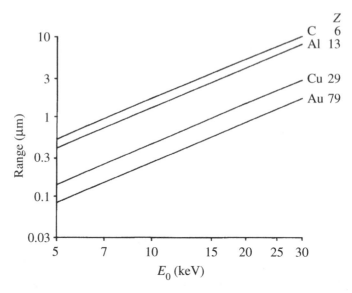

Fig. 2.1. Electron range (defined as the mean straight-line distance from start to finish of electron trajectory in the target) as a function of incident electron energy (E_0) for target elements of various atomic numbers, calculated from the expression of Kanaya and Okayama (1972).

2.2.1 Electron range

The path length of electrons of a given initial energy is inversely proportional to density, and the product of distance travelled and density ('mass penetration') is approximately constant for all elements. The 'range', r, defined as the straight-line distance between the point of entry of an electron and its final resting place, is related to the path length, but is also influenced by elastic scattering (see Section 2.3), which causes the electrons to follow zig-zag paths. The following expression (Kanaya and Okayama, 1972) is useful for obtaining an estimate of r (in micrometres):

$$r = 2.76 \times 10^{-2} A E_0^{1.67} / (\rho Z^{0.89}). \tag{2.1}$$

Ranges for various elements, as a function of E_0, are plotted in Fig. 2.1.

2.3 Elastic scattering

Elastic interactions with atomic nuclei involve large deflections in which there is little transfer of energy, owing to the large mass of the nucleus compared with that of the electron (Fig. 2.2). The angular deflection γ, as derived by Rutherford from classical mechanics, is given by

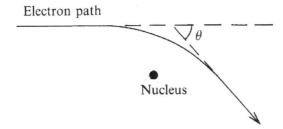

Electron path

Nucleus

Fig. 2.2. Elastic scattering: an incident electron is deflected (without significant energy loss) by the attractive force experienced in passing close to a positively charged nucleus.

$$\tan(\gamma/2) = Z/(1.4pE), \qquad (2.2)$$

where p is the minimum distance (in nanometres) between the undeflected electron path and the nucleus, and is known as the 'impact parameter'. It follows from Eq. (2.2) that elastic scattering is greatest for heavy elements and low electron energies.

Computer simulations of electron trajectories are useful for modelling the spatial distribution of electrons in the target. For this purpose the 'Monte Carlo' method is used: this entails dividing the electron path into short sections and using random numbers to replicate the role of chance in determining the deflection angle (Joy, 1995). Figure 2.3 shows an example of the result of such a simulation.

2.3.1 Backscattering

There is a finite probability of an incident electron being deflected through an angle greater than 90° and emerging from the surface of the target. A similar result can also be obtained as a result of multiple deflections through smaller angles. The fraction of incident electrons which leave the specimen in this way is known as the backscattering coefficient (η) and is strongly dependent on atomic number, because of the increasing probability of high-angle deflection with increasing Z (see Eq. (2.2)). The relationship between η and Z takes the form shown in Fig. 2.4 for normal-incidence electrons (η is greater for oblique angles of incidence). Backscattered electrons have energies ranging up to E_0 (the incident electron energy), and the mean of the energy distribution is highest for elements of high atomic number, for which there is a relatively large probability of high-angle deflections, by comparison with low-Z elements for which multiple low-angle scattering dominates and more energy is lost before the electrons emerge.

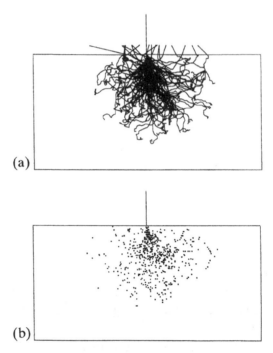

Fig. 2.3. Computer simulations by the Monte Carlo method: (a) electron trajectories and (b) X-ray emission (each dot represents the emission of a photon), for 20-keV incident electrons and a silicon target (side of rectangle $= 3\,\mu$m). (By courtesy of P. Duncumb.)

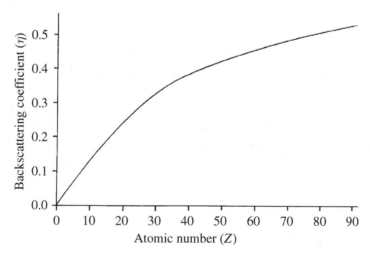

Fig. 2.4. Backscattering coefficient (η) versus atomic number; η is the fraction of incident electrons that leave the target and is almost independent of incident electron energy over the normal range.

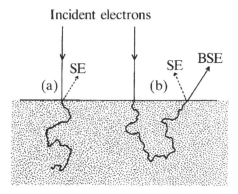

Fig. 2.5. Production of secondary electrons (SE): (a) by incident electrons entering target; and (b) by backscattered electrons (BSE) as they leave.

Elastic scattering causes electrons to travel in different directions after entering the sample and some electrons experience Bragg reflection by atomic planes in crystalline materials. Under normal conditions this has negligible consequences, but, if a suitable experimental arrangement is used, electron backscatter diffraction (EBSD) patterns can be observed (see Section 3.12.3).

2.4 Secondary-electron emission

Electrons originally residing in the specimen that are ejected as 'secondary' electrons are distinguishable from backscattered electrons by their much lower energy, the average being only a few electron volts. Because of their very low energy, only those electrons that originate within a few nanometres of the surface are able to escape. Some of these are produced by incident electrons as they enter the specimen and others by backscattered electrons as they emerge (Fig. 2.5). The secondary electron coefficient, δ, is defined as the number of secondary electrons produced per incident electron. Its value is between 0.1 and 0.2 (approximately), and does not vary smoothly with Z. For incident electron energies below about 5 keV, δ increases because more energy is deposited near the surface, and it is also greater for oblique incidence for the same reason.

2.5 X-ray production

Bombardment of a solid by electrons produces X-rays by two quite distinct mechanisms. A smooth 'continuous' spectrum is produced by electrons interacting with atomic nuclei, whereas the 'characteristic' spectrum contains lines that result from electron transitions between energy levels that are specific to each element.

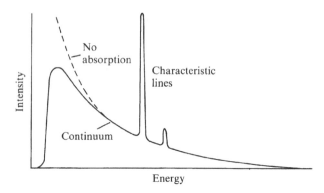

Fig. 2.6. X-ray spectrum plotted against photon energy, showing characteristic lines superimposed on the continuous X-ray spectrum (or 'continuum') produced by incident electrons involved in inelastic collisions with atomic nuclei; the observed spectrum falls off at low energies owing to absorption in the detector window, etc.

2.5.1 *The continuous X-ray spectrum*

When an energetic electron passes through the strong electric field close to an atomic nucleus, it may suffer a quantum jump to a lower energy state, with the emission of an X-ray photon with any energy up to E_0, the initial energy of the electron. The main significance of the resulting continuous X-ray spectrum (also known as the 'continuum' or 'bremsstrahlung') is that it limits the detectability of the characteristic lines of elements present in low concentrations.

The intensity, I, of the continuum can be represented by the following expression (Kramers' law):

$$I = \text{constant} \times Z(E_0 - E)/E, \tag{2.3}$$

where E is the X-ray photon energy and Z is the atomic number (the mean value in the case of a compound). According to Eq. (2.3) the shape of the spectrum (intensity plotted against photon energy) is the same for all elements, while the intensity is proportional to Z. The intensity falls to zero at the 'Duane–Hunt limit', where the X-ray energy is equal to E_0, and rises rapidly with decreasing energy. In observed spectra the intensity falls off at very low energies because of absorption in the detector window and in the sample itself (see Fig. 2.6).

2.5.2 *Characteristic X-ray spectra*

Characteristic X-rays are produced by electron transitions between bound electron orbits, the energies of which are governed primarily by the principal

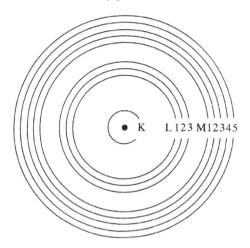

Fig. 2.7. A schematic diagram of inner atomic electron shells; characteristic X-rays are produced by transitions between these shells.

quantum number, n. Inner orbits form closed shells designated K ($n=1$), L ($n=2$), M ($n=3$), etc., in order of increasing distance from the nucleus and decreasing energy (Fig. 2.7). Except for atoms of low atomic number, these shells are normally completely filled and form a 'core' about which outer electrons orbit. The numbers of electrons in the shells are determined by other quantum numbers relating to angular momentum: the K shell contains a maximum of 2, the L shell 8, the M shell 18, etc. The L, M and higher shells are split into subshells with different quantum configurations that result in slightly different energies. The L shell consists of three subshells (L1, L2 and L3), and there are five M subshells. With increasing atomic number the available orbits are progressively filled, those nearest the nucleus (with the highest binding energy) being occupied first.

A necessary condition for the production of a characteristic X-ray photon is the removal of an inner electron, leaving the atom in an ionised state. For a characteristic X-ray line to be produced, the incident electron energy, E_0, must exceed the 'critical excitation energy' (E_c) required to ionise the relevant shell of the element concerned, which varies approximately as Z^2. The probability of ionisation can be expressed as a 'cross-section' (Q), which is small for energies just above E_c, rising to a maximum at about $2E_c$, then falling relatively slowly (Fig. 2.8). It follows that E_0 must be substantially above E_c for there to be reasonably intense characteristic X-ray emission.

The energies of the relevant energy levels can be represented diagrammatically, as in Fig. 2.9: the X-ray photon energy is equal to the difference between the energies of the initial and final levels for the transition concerned. Only

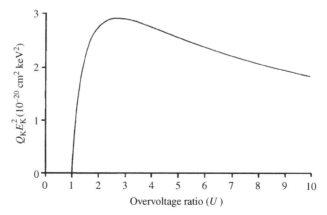

Fig. 2.8. The ionisation cross-section of the K shell (Q_K) expressed as the product $Q_K E_K^2$ (E_K is the K-shell excitation energy), versus the overvoltage ratio $U = E_0/E_K$, where E_0 is the incident electron energy; Q_K represents the probability of the ejection of a K electron, which is a necessary precondition for K X-ray emission.

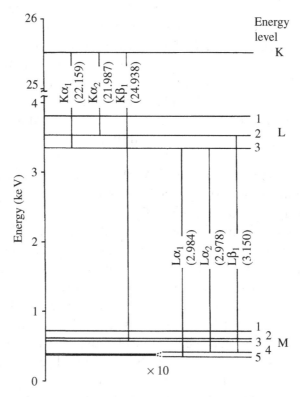

Fig. 2.9. Energy-level diagram for silver ($Z = 47$); the characteristic X-ray energy (given in keV) is equal to the difference in energy between the levels involved in the transition.

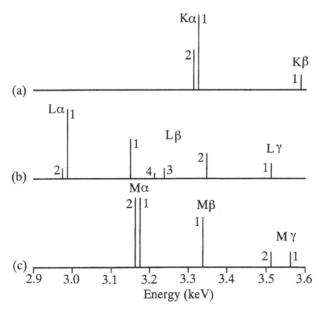

Fig. 2.10. Typical examples of characteristic X-ray spectra (principal lines only are shown): (a) the K spectrum of potassium ($Z = 19$); (b) the L spectrum of silver ($Z = 47$); and (c) the M spectrum of uranium ($Z = 92$).

transitions that obey the rules of quantum theory are allowed: the most significant of these are shown in Fig. 2.9. Lines are designated K, L, etc. according to the shell containing the initial vacancy. Within a given shell, lines in the group which are most intense are labelled α, those in the next most intense group, β, etc., and within each group the lines are numbered in order of intensity (approximately). Typical spectra are shown in Fig. 2.10. (Note that a, b and g are often used instead of α, β and γ on computer screens and print-outs.)

The energy of a given line varies approximately as the square of the atomic number of the emitting element (Moseley's law). The energies and wavelengths[*] of $K\alpha_1$, $L\alpha_1$ and $M\alpha_1$ lines are plotted against Z in Fig. 2.11. For analysis one is mainly concerned with X-ray energies up to about 10 keV, and the $K\alpha_1$ line is used for the analysis of elements of atomic number up to about 30, above which the $L\alpha_1$ line is used (or $M\alpha_1$ for the heaviest elements). Other, less intense, lines are rarely used for analysis, but sometimes interfere with lines that are so used, and therefore cannot be ignored. The relative intensities of these lines depend on the numbers of electrons occupying the energy levels involved and are fairly constant.

---

[*] Wavelength (λ) and energy (E) are related thus: $E\lambda = 12\,398$, in which E is in electron volts and λ is in ångström units ($1\,\text{Å} = 10^{-10}\,\text{m}$).

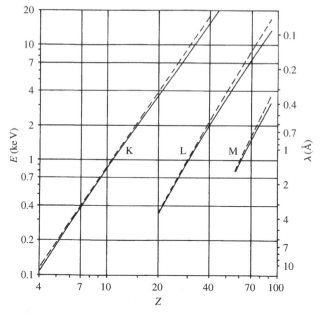

Fig. 2.11. Energy (E) and wavelength (λ) of Kα, Lα and Mα lines (solid lines) and their excitation energies (dashed lines), versus atomic number (Z).

For the most part the relative intensities and positions of X-ray lines are similar for different elements, apart from a shift as a function of atomic number. However, substantial changes occur in L and M spectra for atomic numbers below approximately 26 and 65, respectively, owing to incomplete filling of the relevant shells. As a consequence the number of lines is reduced, also the relative intensity of the α line decreases (disappearing completely in the case of Mα).

2.6 X-ray absorption

The observed intensity of characteristic X-ray lines may be affected significantly by absorption occurring in the sample itself. The effect of absorption in a thin layer is given by the equation

$$I = I_0 \exp(-\mu\rho x), \tag{2.4}$$

where I_0 is the initial intensity, I the intensity after absorption, μ the 'mass absorption coefficient' (cm^2 g^{-1}), ρ the density (g cm^{-3}) and x the path length (cm). Values of μ are widely variable, ranging from less than 100, for X-rays of high energy and absorbers of low atomic number, to more than 10 000, for

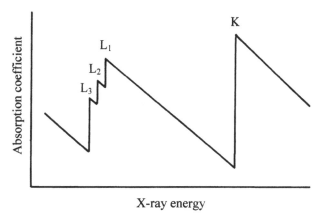

Fig. 2.12. A schematic representation of the variation of the mass absorption coefficient as a function of X-ray energy.

X-rays of low energy and absorbers of high atomic number. In the latter case severe absorption occurs even for x less than 1 μm.

The predominant mechanism of X-ray absorption is the ejection of inner-shell electrons. On a plot of μ versus X-ray energy, sharp discontinuities ('absorption edges') occur at energies corresponding to the critical excitation energies of the different shells of the absorbing element (Fig. 2.12). Below such an edge the X-rays have insufficient energy to ionise the shell concerned.

2.7 The Auger effect and fluorescence yield

The energy released when an atom ionised in an inner shell returns to its normal state by electron transitions from outer levels may be used to eject another bound electron instead of an X-ray photon (the 'Auger effect'). The main significance of this is its influence on X-ray intensities, but also it is the basis of 'Auger analysis' (Section 1.4.4). The 'fluorescence yield', denoted by ω_K (for the K shell), is the probability of ionisation being followed by the emission of an X-ray photon rather than an Auger electron. This increases rapidly with increasing Z (Fig. 2.13), which tends to compensate for the decrease in Q_K with Z already noted. The fluorescence yields of other shells behave similarly.

2.8 Cathodoluminescence

In some types of sample, electron bombardment stimulates the emission of light by the process of cathodoluminescence (CL). In a non-metallic material

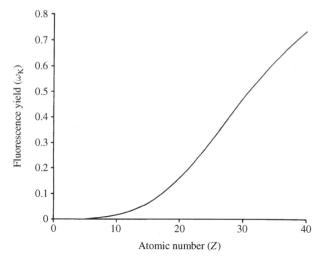

Fig. 2.13. The atomic-number dependence of the 'fluorescence yield' of the K shell (ω_K), which is the probability of K-shell ionisation being followed by characteristic X-ray emission (rather than Auger electron emission).

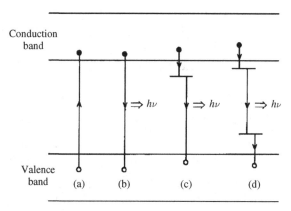

Fig. 2.14. Cathodoluminescence: (a) electron bombardment raises an electron from the valence to the conduction band; de-excitation may occur directly (b) or via localised levels in the band gap, associated with defects or impurities (c) and (d). The photon energy $h\nu$ is the difference in energy between the initial and final levels.

incident electrons cause excitation by raising electrons in the valence band to the normally empty conduction band, from which they return to their original state in one or more steps. The surplus energy can be dissipated in various ways, one of which is the emission of photons (Fig. 2.14). Only relatively low energies (a few electron volts) are involved, and the wavelengths fall within the visible region (sometimes extending into the ultra-violet or infra-red).

Minerals that exhibit CL include diamond, quartz, corundum, rutile, cassiterite, benitoite, willemite, halite, fluorite, spinel, calcite, dolomite, kaolinite, apatite, barite, strontianite, sphalerite, zircon, feldspar, jadeite, diopside, wollastonite, forsterite and fayalite.

Localised energy levels in the gap between valence and conduction bands, arising from lattice defects, interstitial ions, or substitutional impurity atoms, often play an important role in CL emission (see Fig. 2.14). Certain elements behave as 'activators', small concentrations of which are sufficient to produce CL. Others (notably divalent Fe) have the effect of 'quenching' CL emission. The intensity of some forms of CL emission is strongly influenced by the density of defects, which is dependent on factors such as temperature of formation, cooling rate, deformation and irradiation. High defect densities, however, may suppress CL by promoting alternative modes of de-excitation.

The colour of CL emission depends on the difference in energy between the states concerned. Most commonly the energy is not narrowly defined and the emission takes the form of a band (Fig. 2.15(a)). In a few cases line spectra characteristic of the impurity element are produced (Fig. 2.15(b)).

Excitation of CL is not very sensitive to the beam accelerating voltage, but sometimes it is advantageous to use a relatively high value (at least 20 kV) because this enables the electrons to penetrate the non-luminescent damaged surface layer. Prolonged electron bombardment tends to cause fading, whereby the intensity declines, sometimes irreversibly. Cooling the specimen below room temperature significantly increases CL intensity for certain types of sample.

Cathodoluminescence can be detected in SEMs and EMPs with appropriate light-detection equipment (Section 3.12.2). Owing to the complexity of the factors governing CL emission in natural materials, unambiguous elemental analysis is problematic, but nevertheless information not easily obtained by other means in a range of geological applications can be obtained from CL images (see Section 4.8.4).

2.9 Specimen heating

The fraction of the energy in the incident electron beam which reappears in the form of X-rays, light, etc. is small, most being converted into heat in the target. The temperature rise ΔT can be estimated from the following expression:

$$\Delta T = 4.8E_0 i/(kd), \qquad (2.5)$$

where E_0 is the incident electron energy (keV), i the current (μA), k the thermal conductivity (W cm^{-1}K^{-1}) and d the beam diameter (μm). For metals k is

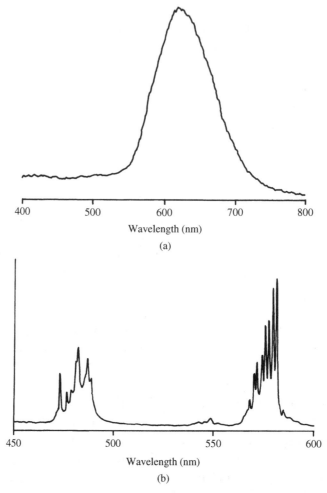

Fig. 2.15. Cathodoluminescence spectra: (a) the emission band caused by Mn in carbonates etc.; and (b) emission lines caused by rare earths (mainly Dy) in zircon.

typically in the range 1–4 and under normal conditions ΔT is negligible. On the other hand, for materials of low thermal conductivity, including many minerals, the temperature may be raised significantly: for example, in the case of mica $(k = 6 \times 10^{-3})$ the calculated temperature rise is 160 K for $E_0 = 20$ keV, $i = 10$ nA and $d = 1$ μm. The heating effect can be moderated by reducing i or increasing d, or by using a surface coating of a good conductor (see Section 8.6).

3

Instrumentation

3.1 Introduction

Scanning electron microscopes and electron microprobes have much in common, including a source of electrons (an 'electron gun'), and electron lenses to focus the beam on the specimen, which together form the 'column'. Column design is similar in principle for both types of instrument: descriptions in the following sections therefore apply equally to both for the most part (differences are explained where necessary). Beam-deflection coils and electron detectors enable scanning images to be produced. X-ray spectrometers (described in the next chapter) are a common SEM accessory and an essential component of the EMP. Other types of detector are available optionally. Features also described in the following sections include the vacuum system, specimen stage and, in the case of EMPs (but not usually SEMs), optical microscope.

3.2 The electron gun

The source of electrons in EMP and SEM instruments is held at a negative potential (typically 10–30 kV), which accelerates the electrons towards the sample. The commonest type of emitter is a tungsten filament (about 0.1 mm in diameter) bent into a 'hairpin' shape and attached to legs mounted on an insulator (Fig. 3.1). This is heated electrically to about 2700 K, giving electrons sufficient thermal energy to overcome the potential barrier at the surface. The grid or 'wehnelt' (Fig. 3.2) is held at a negative potential relative to the cathode and limits the effective emitting area of the filament to the region close to the tip. The anode consists of an earthed plate with an aperture to let the beam pass.

The lifetime of the filament is governed by thinning due to evaporation of tungsten and is strongly temperature-dependent. A filament should last for several

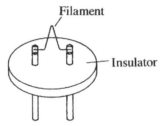

Fig. 3.1. Tungsten 'hairpin' filament used as the electron source in SEMs etc.

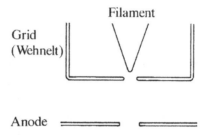

Fig. 3.2. Electron gun (triode type): electrons emitted by the filament are accelerated towards the anode; the grid (or 'wehnelt') controls emission from the filament tip.

weeks if correctly adjusted and operated. When fitting a new filament it should be accurately centred in the wehnelt aperture and set to the correct height.

With increasing filament heating current the emission current rises rapidly at first, but then levels out to a 'saturated' value of 50–100 μA. The small fraction of this current that reaches the specimen behaves similarly, but reaches a 'plateau' at a somewhat higher filament temperature (Fig. 3.3). In the unsaturated condition, electrons from the sides of the filament as well as from the tip contribute to the beam, and the focal spot is enlarged. It is therefore important to operate on the plateau in order to achieve both a small beam diameter and insensitivity to variations in temperature. The current meter fitted as standard in EMP instruments may be used for finding the 'plateau'; SEMs usually lack such a meter, but saturation can be determined by observing the brightness of the image. Operating significantly above the 'knee' of the curve has no advantages and results in reduced filament life. The saturation temperature can be lowered, and hence filament life increased, by increasing the distance between the tip of the filament and the wehnelt (the maximum current obtainable is reduced, but for most purposes this is unimportant).

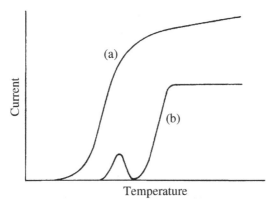

Fig. 3.3. The relationship between the filament temperature and the current (a) emitted by the electron gun and (b) reaching the specimen; the normal operating point is just above the 'knee' of the latter curve.

3.2.1 High-brightness electron sources

Replacing the tungsten filament by lanthanum or cerium hexaboride (LaB_6 or CeB_6) gives an order-of-magnitude increase in brightness (and a corresponding increase in current for a beam of given diameter), but higher vacuum is required. In a 'field emission' (FE) source, electrons are drawn from an extremely fine tungsten point by a strong electric field. Field emitters are very sensitive to vacuum conditions and require ultra-high vacuum. They can be used at room temperature, but operation at an elevated temperature has the advantage of minimising gas adsorption, hence giving more stable emission. The 'Schottky' type of field emitter has a tip of larger radius, which is coated to enhance electron emission. Its somewhat lower brightness is compensated by higher maximum beam current and less extreme vacuum requirement. Field emission sources offer the best available performance for high-resolution SEM, but their higher cost is less justifiable for X-ray analysis and other modes of operation, for which relatively high beam current is desirable and resolution is limited mainly by spreading in the sample rather than by beam diameter.

3.3 Electron lenses

The effective source diameter obtained with the conventional (tungsten-filament) electron gun is about 50 μm. Magnetic electron lenses, each consisting of a coil of copper wire carrying a direct current, surrounded by an iron shroud, project a demagnified image of the source onto the surface of the specimen. The

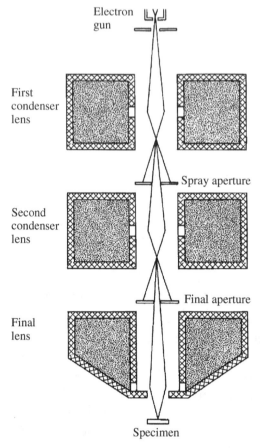

Fig. 3.4. Schematic diagram of a three-lens probe-forming column as used in SEMs etc.: lenses produce a demagnified image of the electron source; aperture diaphragms intercept the unwanted part of the beam.

electromagnetic field is contained within the iron except where there is a gap, in the region of which the field on the axis rises sharply to a peak. Interaction with this field causes electrons to be deflected towards the axis, giving properties analogous to those of convex glass lenses used for focussing light. The strength of the lens can be controlled by varying the current in the coil.

For a single lens the demagnification factor equals the source-to-lens distance divided by the lens-to-specimen distance. To obtain sufficient overall demagnification, several (typically three) lenses are used (Fig. 3.4). (Fewer lenses are required with a field emission source, in view of the small source diameter.) The first two lenses in such a system are usually known as 'condensers'. In SEMs the final ('objective') lens is commonly of the 'pinhole' type (as shown in Fig. 3.4), in which the lower polepiece has a small internal

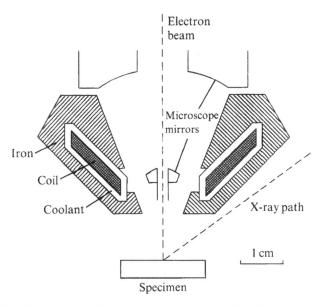

Fig. 3.5. Final lens design for an electron microprobe (by courtesy of JEOL Ltd): liquid-cooled miniature lens allows space for optical microscope components and X-ray paths to spectrometers.

diameter, which minimises the magnetic field in the region of the specimen. A smaller focal spot can be achieved with an 'immersion' lens, in which the bore of the polepiece is large enough to allow a small specimen to be raised into the bore, where it is immersed in the magnetic field of the lens. Such a lens can be operated with a very short focal length, thereby minimising aberrations (see the next section). It is harder to collect electron and other signals, but this drawback is avoided in the 'snorkel' lens, in which the focussing field extends below the polepiece, and short focal length can be combined with larger specimens and easier access.

Somewhat different design criteria apply in the case of the electron microprobe, in which an optical microscope and multiple WD spectrometers must be accommodated, and a very small beam diameter is not necessary. Figure. 3.5 shows a 'mini-lens' used in an EMP: this employs a coil of small dimensions carrying a relatively high current, allowing maximum space for other components.

3.3.1 Aberrations

The strength of a magnetic electron lens increases with the distance of the electron trajectory from the axis, giving rise to 'spherical aberration' analogous to that of a simple biconvex glass lens (Fig. 3.6). This defect is important only in the final lens, and can be controlled by means of a limiting aperture.

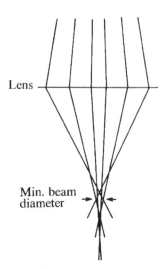

Fig. 3.6. Spherical aberration: the lens focusses outer rays more strongly than those close to the axis, resulting in enlargement of the beam diameter at the focus.

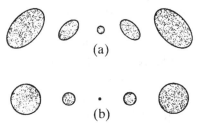

Fig. 3.7. The appearance of the electron beam in a through-focus series: (a) with and (b) without astigmatism (this behaviour can be observed on a cathodoluminescent sample).

Also, it decreases with decreasing focal length, and for high-resolution SEM imaging may therefore be minimised by decreasing the 'working distance' between specimen and lens (this is not applicable to EMP instruments, in which this distance is fixed).

Another important aberration is astigmatism caused by small lens imperfections or contamination on apertures, the effect of which is that two elliptical foci (with perpendicular axes) occur in separate planes. The result is loss of resolution in SEM images. This can be corrected by means of a 'stigmator' consisting of coils that create astigmatism that cancels out that already present. In EMP instruments, the stigmator can be adjusted by observing the change in shape of the spot of light produced by the beam on a cathodoluminescent sample as the lens is swept through focus (Fig. 3.7). Alternatively, the adjustment can be made while observing a scanning image and varying the stigmator controls, which is the method used in SEMs.

3.3.2 *Apertures*

An aperture limiting the beam diameter in the final lens is essential to control spherical aberration, as noted in the previous section. This consists of a thin disc (diaphragm) with a central hole allowing the beam to pass through, usually made of a metal such as platinum or molybdenum (Fig. 3.4). Several interchangeable apertures of different sizes may be provided: for high spatial resolution a small aperture should be selected, but, when a larger beam diameter is acceptable, more current can be obtained with a larger aperture. Additional 'spray apertures' located between the electron gun and the final lens intercept the outer parts of the beam which might otherwise be scattered on striking lens bores etc. When astigmatism caused by contamination becomes too much to be corrected by the stigmator, the relevant aperture must be either cleaned or replaced.

3.4 Beam diameter and current

As described in the preceding sections, the beam diameter in the specimen plane is determined principally by the effective source size, demagnification by the electron lenses and spherical aberration. Demagnification is determined by the condenser lens settings (sometimes labelled 'spot size'). On increasing the condenser strength in order to reduce the beam diameter, the divergence of the beam increases and the fraction passing through the final aperture decreases. Also, the reduction in aperture size necessary in order to minimise spherical aberration entails a further decrease in current. As a result of these factors the maximum current obtainable varies with beam diameter, d, approximately as $d^{8/3}$ (Fig. 3.8). More current can be obtained in a beam of given diameter by decreasing the working distance (possible only in SEMs) so that spherical aberration is reduced (see Section 3.3.1). The minimum beam diameter attainable with a high-performance SEM is typically about 2 nm, but is larger for EMPs.

3.5 Column alignment

The direction of the electron beam emerging from the gun is sensitive to the position of the filament relative to the wehnelt aperture and not only is it impossible to set this initially with perfect precision, but also the filament tip is

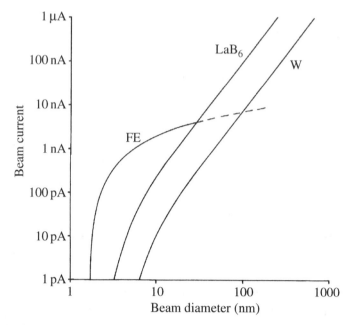

Fig. 3.8. Beam current versus diameter for a typical instrument, using different types of electron source: W, tungsten filament; LaB_6, lanthanum hexaboride; and FE, field emission.

liable to wander in use. In addition, the components of the electron optical column (electron gun, lenses etc.) are subject to small misalignments. Adjustment of beam alignment can be carried out by means of magnetic coils (in place of mechanical centring adjustments) that 'steer' the beam along the correct path. In a computer-controlled instrument the adjustment can be done automatically, and optimum settings (which vary with accelerating voltage) can be saved in a 'set-up' file.

Usually the final lens aperture is centred mechanically, by finding the position for which there is no lateral movement of the beam as the lens focus is varied. A useful aid is a 'wobbler', which oscillates the focus setting while the operator adjusts the aperture position.

3.6 Beam current monitoring

It is desirable (especially for X-ray analysis) to have means for monitoring beam current. The current flowing from specimen to earth does not give a true indication of the current in the incident beam because secondary and back-scattered electrons leaving the sample reduce the apparent current by a variable amount. It is therefore preferable to collect the current in a 'Faraday cup' (or 'cage') consisting of a deep recess in a solid conducting block, preferably made of a material of low atomic number (e.g. carbon) to minimise back-scattering. This may be mounted on the specimen holder or, more conveniently, on a retractable arm (a standard feature of EMPs, but not SEMs).

As noted in Section 3.4, there is a direct relationship between the strength of the condenser lenses and the beam current. The required current may therefore be obtained by adjustment of the condenser lenses, which, in a computer-controlled instrument with current monitoring, can be done automatically by means of a software command. (This also has the effect of changing the final beam diameter.)

Drift in beam current as a function of time is caused mainly by movement of the tip of the filament, which can be corrected using beam alignment coils (as discussed in the previous section). Such drift is a potential source of error in automated quantitative X-ray analysis sessions, which may extend over several hours. One solution to this problem is to insert the Faraday cup before each measurement and normalise the intensities. A better alternative, however, is to use a regulating system that continuously adjusts the condenser lenses to maintain a constant current. This requires a double aperture, in which part of the beam passing through the first aperture is intercepted by the second (smaller) one, and the current collected by this is used as input to a feedback amplifier. (It is assumed that this current is proportional to that passing through the second aperture.) This is a normal feature of EMPs but not SEMs.

3.7 Beam scanning

Scanning images are produced by sweeping the beam across the sample in a television-like 'raster' while displaying the output of an electron detector on the screen of a synchronously scanned visual display unit (VDU). The electron beam is deflected by means of coils located above the final lens, which enable the beam to be 'pivoted' about the centre of the lens, as shown in Fig. 3.9. These are supplied with a 'sawtooth' current waveform derived from line and frame scan generators. The ratio of frame and line scan frequencies determines the number of lines in the image (typically 500–2000).

Instead of scanning a rectangular raster, the beam can be swept along a single line by using only one set of coils, in order to produce a line plot, which is useful for some purposes. The deflection system can also be used to move the beam around in 'spot' mode for X-ray or other forms of analysis on selected points.

'Analogue' scanning systems have been superseded by digital systems in which the beam deflection is computer-controlled via a digital-to-analogue converter (DAC) and the output from the signal detector is converted into

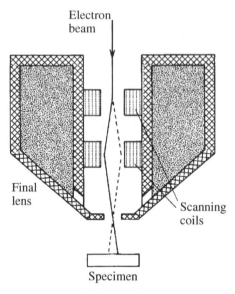

Fig. 3.9. Electron-beam scanning using double deflection coils in the bore of the final lens.

digital form by an analogue-to-digital converter (ADC), so that the intensity at each point (or 'pixel') in the image is recorded as a number. Such images are displayed on a computer monitor and can be saved on the hard disc of the computer, from which they can be printed out, copied to other media for long-term storage, or transferred to another computer via a network connection.

For scanning large areas, stage movement rather than beam deflection can be used. However, owing to the relatively slow speed of the mechanical stage movements, this entails a long acquisition time, which is appropriate for X-ray 'maps' used to show element distributions, but less so for electron images.

3.8 The specimen stage

Specimens for the SEM are commonly fixed to a 'stub' consisting of a metal disc with a projecting peg (Fig. 3.10). This is mounted on the stage mechanism, which incorporates linear movements in the x and y directions (perpendicular to the column axis), enabling different areas of the specimen to be imaged and/or analysed, and in the z direction (parallel to the axis), which serves to locate the surface of the specimen at the required height relative to the final electron lens (also light microscope and X-ray spectrometers, if fitted). Tilt and rotation movements allow adjustment of the orientation of the specimen relative to the beam and electron (or other) detectors in order to optimise the

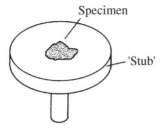

Fig. 3.10. 'Stub' used for mounting SEM specimens.

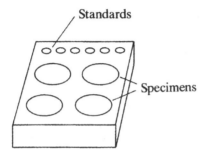

Fig. 3.11. Typical specimen holder for EMPA, holding a number of small standards and several specimens (round in this case, but they may be rectangular).

image. It is desirable for these motions to be 'eucentric' (centred about the column axis).

For quantitative X-ray analysis, constant geometry is required, ideally with normal electron incidence, and the arrangements used in EMPs are therefore different. A typical EMP specimen holder is illustrated in Fig. 3.11. Several specimens and standards may be mounted together, with their front surfaces lying in a common plane defined by a ledge at the front of the holder (described as 'front referencing', as distinct from 'back referencing' of SEM specimens mounted as described above). The holder is attached to a mechanism that provides orthogonal movements, which in modern instruments are computer-controlled. The minimum step size is typically 1 μm or less, and 'micro-stepping' is sometimes available to give ultra-fine control of position.

Optional specialised facilities include cooling (using either a thermo-electric device or liquid nitrogen), which enables aqueous samples to be examined in a frozen state, and heating, enabling the response of mineral samples etc. to be viewed 'live' (for example, see Kloprogge, Boström and Weier (2004)). The latter is most suitable for use in an environmental SEM (Section 3.10.2), owing to the emission of gases during heating.

3.9 The optical microscope

Optical microscopes in current EMP instruments use a reflecting objective lens mounted coaxially with the electron beam, as shown in Fig. 3.12 (though other configurations have been used in the past). This enables the sample to be viewed at normal incidence while under electron bombardment, so that cathodo-luminescence (Section 2.8) can be used as an aid in determining the location of the beam, on a suitable target such as benitoite, willemite, periclase, glassy carbon, or microscope-slide glass.

The focal plane of the microscope is fixed, and the specimen is moved in the vertical (z) direction to obtain a sharp image, ensuring that the surface is always in the same plane, which is essential for WD (though not ED) analysis. In some instruments automatic focussing is available. Usually a CCTV camera is fitted in place of the eyepiece to provide an image on a video screen or computer monitor.

In instruments with fixed magnification this is fairly high, as required for precise location of points for analysis. Sometimes there is provision for 'zoom-ing' to low magnification, which is useful when searching specimens for areas of interest. Illumination by transmitted light, with polariser and analyser, is an optional feature that is highly desirable for geological applications.

Some SEMs have a low-power microscope using oblique incidence, but most have none. This is a handicap for geological applications, though back-scattered electron images serve as a partial substitute.

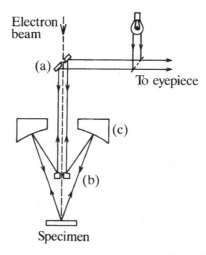

Fig. 3.12. Reflecting microscope mounted coaxially with the electron beam, as fitted to electron microprobe instruments.

3.10 Vacuum systems

Electron beam instruments must be evacuated sufficiently well to avoid damage to the electron source and high-voltage breakdown in the gun, as well as allowing electrons to reach the specimen without being scattered. Taking these considerations into account, it is desirable for the operating pressure to be below 10^{-5} mbar*, though a lower pressure in the electron gun is required when using a high-brightness electron source (Section 3.2.1).

A simple vacuum system employing a mechanical rotary pump and an oil diffusion pump is shown schematically in Fig. 3.13. To pump the chamber from atmospheric pressure, V2 is opened (with V1 and V3 closed). When a pressure of about 0.1 torr is attained with the rotary pump, V2 is closed and both V1 and V3 are opened, bringing into operation the diffusion pump (which must be backed by the rotary pump whilst in use). In contemporary instruments the vacuum system is computer-controlled, so that the required operations are executed automatically and inappropriate actions that could cause damage are barred. Pump-down time is minimised by venting with dry nitrogen when changing specimens, etc., to avoid introducing water vapour. For some purposes better vacuum than that provided by the oil diffusion pump is needed (for example when a field emission source is used), in which case a turbomolecular or ion pump may be used.

The following additional features are sometimes provided (in EMPs more often than in SEMs): (1) a gun-isolation valve to enable the gun to be vented and pumped independently for filament replacement; (2) a specimen airlock to enable specimens to be changed without venting the whole column;

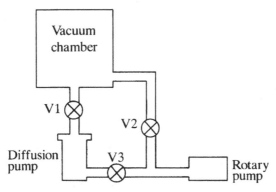

Fig. 3.13. A simple vacuum system with diffusion and rotary pumps (see the text for details of operation).

---

* Units of pressure are the torr (1 mm Hg), the millibar (10^{-3} atm) and the pascal (1 N m^{-2}), which are related as follows: 1 torr = 1.3 mbar = 130 Pa.

and (3) a separate chamber for WD spectrometers, allowing them to be evacuated by rotary pump only.

3.10.1 Contamination

Residual hydrocarbon molecules adsorbed on exposed surfaces in the vacuum chamber are polymerised by electrons, leaving a carbon deposit. This causes contamination of apertures and other components, which can lead to instability due to charging. Carbon is also deposited on the specimen at the point of impact of the beam. In many applications this is not very important, but absorption of long-wavelength X-rays passing through the contamination layer can be significant. The following measures may be applied to minimise contamination: (1) replacing the diffusion pump by an oil-free type of high-vacuum pump (see above); (2) fitting a vapour trap in the rotary pump backing line; (3) replacing the rotary pump by an oil-free type; (4) using a liquid-nitrogen cold trap above the pump and/or close to the specimen; and (5) introducing a jet of air or oxygen via a fine capillary tube close to the specimen. The cold trap should be warmed up before venting, to avoid condensation of water. In the case of the gas jet, the flow rate must be adjusted so that the pressure rise in the chamber is within acceptable bounds.

3.10.2 Low-vacuum or environmental SEM

Instead of the SEM specimen chamber being maintained at the same low pressure as the column, a relatively high pressure is advantageous for some purposes. This is achieved by a valve leaking gas into the specimen chamber, with differential pumping between the specimen chamber and the upper sections of the column. An instrument designed to operate in this mode is known as a 'low-vacuum' or 'environmental' SEM (LVSEM or ESEM). The term 'variable-pressure SEM' (VPSEM) refers to the ability to function in either normal or low-vacuum mode.

The main advantage of low vacuum is that specimen charging is neutralised by positively ionised gas atoms, making coating of insulators unnecessary. For this purpose, a pressure of 0.1 mbar is sufficient. With a pressure of about 20 mbar, liquid water is stable and wet samples can be used (this is of interest mainly for biological applications). Under low-vacuum conditions, normal backscattered-electron detectors can be employed, but special arrangements are necessary for SE imaging (see Section 3.11.1). The relatively poor vacuum does not seriously interfere with X-ray detection, but spatial resolution is degraded by scattering of the electrons in the beam (for further details, see Danilatos (1994)).

3.11 Electron detectors

The commonest type of SEM image is derived from secondary electrons ejected from the sample by the incident electrons (Section 2.4). In most cases a detector for backscattered as well as secondary electrons (Section 2.3.1) is also provided. Both modes of detection are also available in EMPs.

3.11.1 Secondary-electron detectors

Usually secondary electrons are detected by means of a 'scintillator', which produces light when bombarded with electrons, the light being converted into an electrical signal by a photomultiplier. However, secondary electrons are emitted with energies of only a few electron volts and must be accelerated to produce a reasonable output from the scintillator. A positive potential (e.g. 10 kV) is therefore applied to a thin metal coating on the scintillator.

The detector most commonly used in SEMs is the Everhart–Thornley (E–T) type illustrated in Fig. 3.14. This has a mesh in front of the scintillator, which can be biassed to control electron collection. With a positive bias (e.g. 200 V), secondary electrons are attracted and, after passing through the spaces in the mesh, are accelerated towards the scintillator.

When the specimen is immersed in the magnetic field of the final lens in order to achieve the highest possible spatial resolution (see Section 3.3),

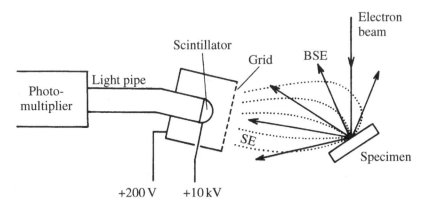

Fig. 3.14. Everhart–Thornley detector, as used in SEMs: low-energy secondary electrons (SE) are attracted by +200 V on the grid and accelerated onto the scintillator by +10 kV; light produced by the scintillator passes along a transparent 'light pipe' to an external photomultiplier, which converts light into an electrical signal; backscattered electrons (BSE) are also detected, but less efficiently because they have higher energy and are not significantly deflected by the grid potential.

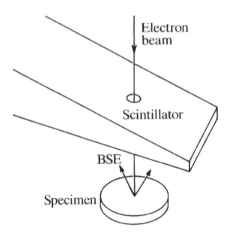

Fig. 3.15. Backscattered-electron scintillation (Robinson) detector.

secondary electrons are trapped in the field and are unable to reach a detector of the conventional type. To overcome this problem a detector is mounted inside the column.

A special type of detector is also required in the low-vacuum or environmental SEM (Section 3.10.2), since secondary electrons cannot move freely in the relatively high gas pressure prevailing in the specimen chamber. For this purpose an insulated plate is mounted on the final lens polepiece and held at a positive potential of around 1 kV. Secondary electrons are detected by means of the signal appearing at the cathode owing to ionisation in the gas.

3.11.2 *Backscattered-electron detectors*

If a negative voltage is applied to the mesh of an E–T detector, secondary electrons are repelled and only backscattered electrons, which are unaffected by the bias owing to their much higher energy, are detected. However, the efficiency in this mode is low because of the small solid angle subtended by the detector.

A more efficient alternative is the 'Robinson' detector, which consists of a scintillator located immediately above the specimen, usually mounted on a retractable arm, with a hole to allow the beam to pass (Fig. 3.15). The large solid angle enables relatively noise-free BSE images to be obtained.

High efficiency can also be obtained with a solid-state detector mounted coaxially with the beam directly above the specimen. These are often divided into sectors (Fig. 3.16), enabling different types of signal to be produced by combining the outputs of the sectors in different ways, and are commonly mounted on a retractable arm. An alternative arrangement used in some

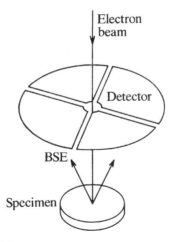

Fig. 3.16. Annular solid-state backscattered-electron detector divided into sectors allowing discrimination between topographic and compositional contrast.

electron microprobes is to dispose a number of individual detectors around the specimen between the spectrometer ports, the signals from which are combined. Solid-state detectors have a slower response than scintillators and are less suitable for use in fast scanning mode. Also, they are usually inefficient for electron energies below about 10 keV, and are sensitive to light from the microscope lamp (if any).

3.12 Detection of other types of signal

Various other types of signal can be used in SEM and EMP instruments, the most important being X-rays (see the following chapter). Some others are covered in the following sections.

3.12.1 Auger electrons

Auger electrons are emitted with energies (mostly in the range 0–3 keV) characteristic of the element concerned (Section 2.7). In order to detect them, an electron spectrometer is required. Usually this employs a cylindrical electrostatic mirror: the energy of the electrons reaching the detector via the exit slit is determined by the potential of the mirror, which is swept in order to produce a spectrum. Scanning images showing elemental distributions are obtained by setting the spectrometer on a particular line.

Auger analysis requires a very clean specimen surface, so ultra-high vacuum (e.g. 10^{-10} torr) is necessary. Ordinary SEMs do not satisfy this requirement, hence a purpose-built 'scanning Auger microprobe' (SAM) is preferable. High Auger-electron collection efficiency can be obtained with a cylindrical mirror spectrometer mounted directly above the specimen.

3.12.2 Cathodoluminescence

Cathodoluminescence (CL) from specimens under electron bombardment (Section 2.8) can be observed in an electron microprobe through the optical microscope. By defocussing or scanning the beam, CL emission from a large area (limited by the field of view) may be observed. Images of larger areas may be obtained by scanning the specimen stage, with a focussed beam.

In SEMs the light can be detected with a photomultiplier (PM) mounted on a window in the specimen chamber, and scanning CL images of any magnification (within the available range) can be produced by modulating the display with the PM output. The light may be focussed onto the PM entrance window with a lens mounted in the specimen chamber. By inserting colour filters different wavelength bands can be selected; 'real' colour images can then be reconstructed by combining red, green and blue images.

Higher sensitivity can be obtained by using a retractable ellipsoidal or paraboloidal mirror with a hole to allow the beam to pass (Fig. 3.17). With the point of impact of the beam at the focus of the mirror the collection

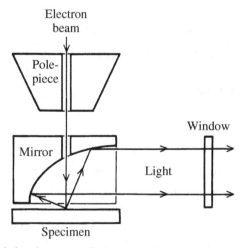

Fig. 3.17. Cathodoluminescence light-collection system with a high-efficiency paraboloidal mirror focussing light into a parallel beam for detection by an externally mounted photomultiplier.

efficiency is very high for a small area (typically of the order of 100 µm diameter); larger areas can be covered by defocussing the mirror, at the expense of reducing its efficiency. This form of light collection is often used in conjunction with a grating spectrograph, in order to obtain adequate intensity. Movable mirrors allow operation in either 'panchromatic' mode (detection of light of all wavelengths) or 'monochromatic' mode (collection of light of a narrow band of wavelengths as selected by the spectrograph). By using different diffraction gratings and detectors, wavelengths from ultra-violet to infra-red can be recorded. Resolution is controlled by the widths of the entrance and exit slits (high resolution entails a sacrifice in intensity). By placing a CCD camera in the focal plane of the spectrograph a range of wavelengths can be recorded in 'parallel' mode, enabling complete spectra to be recorded at each point in a scanned raster.

Cathodoluminescence emission can also be observed with an optical micro-scope fitted with a relatively simple device in which the specimen is bombarded with a broad beam of electrons. However, using an electron microprobe or SEM equipped with a CL detector as described above enables a lower current to be used, with less risk of damage to the specimen; also higher resolution and magnification are available. In addition, weaker CL emission and a wider range of wavelengths (extending beyond the visible region) can be detected, though the capabilities of the CL microscope can be enhanced by using a sensitive CCD camera in place of film.

For a discussion of applications of CL, see Section 4.8.4.

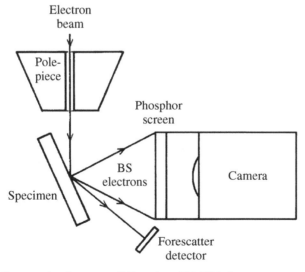

Fig. 3.18. Electron backscatter diffraction (EBSD) detector.

3.12.3 Electron backscatter diffraction

The necessary additional equipment for EBSD is available as an SEM accessory. The specimen is steeply tilted in order to obtain optimum diffraction
conditions and the pattern appears on a phosphor screen that converts electrons to visible photons, and is recorded with a camera (Fig. 3.18). For fast
recording, a normal video camera can be used, but for maximum sensitivity a
slow-scan cooled CCD camera is required. Often a 'forescatter' detector is
mounted below the camera, providing an alternative means for recording
images. The detector assembly is normally retractable to allow normal SEM
operation.

Applications of EBSD are discussed in Section 4.8.3.

4

Scanning electron microscopy

4.1 Introduction

The scanning electron microscope consists essentially of the following (as described in Chapter 3): a source of electrons; lenses for focussing them to a fine beam; facilities for sweeping the beam in a raster; arrangements for detecting electrons (and possibly other signals) emitted by the specimen; and an image-display system. Secondary-electron (SE) images, which show topographic features of the specimen, are the most commonly used type. Backscattered-electron (BSE) images are principally used to reveal compositional variations. An X-ray spectrometer (Chapter 5) is an optional extra enabling the SEM to be used for element mapping and analysis (Chapters 6–8). Other types of image can also be produced, as described at the end of this chapter.

4.2 Magnification and resolution

The magnification of a scanning image is equal to the ratio of the size of the image as viewed by the user to that of the raster scanned by the beam on the specimen. The minimum magnification is determined by the maximum angle through which the beam can be deflected, and depends on the working distance, being least when this is greatest. Typical minimum magnification is about 10, with a scanned area of the order of $1 \, cm^2$. Magnification can be increased by reducing the amplitude of the scanning waveform. There is no advantage in increasing magnification beyond the point at which the image starts to appear unsharp. The useful maximum is thus related to resolution, and for most purposes is in the range 10^4–10^6, according to the type of image, the nature of the specimen and the operating conditions. The ability to 'zoom' over a wide range is illustrated in Fig. 4.1.

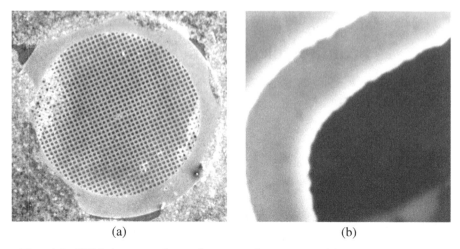

(a)   (b)

Fig. 4.1. SEM images of an electron microscope grid (3 mm diameter); magnification (a) 10 and (b) 10 000.

Resolution, defined as the size of the smallest detail clearly visible in the image, is limited not only by the diameter of the electron beam but also by the interaction between the electrons and the specimen. Beam diameter is determined by various instrumental factors, as discussed in Section 3.4, and can be reduced to a few nanometres in principle. In many applications, however, the ultimate resolution is not needed and a larger beam diameter can be used, with the advantage that more current is then available. The resolution limit as determined by beam–specimen interaction ranges from approximately 1 μm in X-ray images to less than 10 nm for SE images (under favourable circumstances). In digital images the pixel size limits the maximum attainable resolution.

4.3 Focussing

Correct focus setting is achieved by adjusting the control to obtain the sharpest possible image of fine detail in the specimen, preferably with the magnification set to a high value. For this purpose fast (TV rate) scanning is desirable. The beam is perfectly focussed only in a single plane. However, the depth of focus is large compared with that of an optical microscope, owing to the small convergence angle of the beam, so sharpness is often adequate for all parts of the specimen. For tilted specimens, dynamic focus correction linked to deflection can be used.

4.3.1 *Working distance*

A short working distance is optimal for high-resolution SE imaging but is incompatible with some other types of image (for example, an X-ray detector may be unable to 'see' the sample when the working distance is less than a certain value). The working distance can be set by adjusting the final lens focus to the required setting, then moving the specimen stage vertically until the image is sharp. The focus setting depends on the accelerating voltage, and must be adjusted when this is changed.

4.4 Topographic images

The principal function of the SEM is to produce images of three-dimensional objects (the main advantages over the optical microscope are greater depth of focus and higher resolution). Usually secondary-electron images are employed to show topographic contrast, but backscattered-electron images, though mostly used to show compositional differences, may also contain topographic information.

4.4.1 Secondary-electron images

Secondary electrons are emitted from very near the surface of the sample, with energies of a few electron volts (Section 2.4). The SE yield increases with decreasing angle between beam and specimen surface (Fig. 4.2), giving a result that is very similar to that produced by illuminating a solid object with partly directional and partly diffuse light, making the topographic inform-ation contained in such images easy to assimilate intuitively (Fig. 4.3(a)). The Everhart–Thornley type of electron detector (Section 3.11.1) attracts

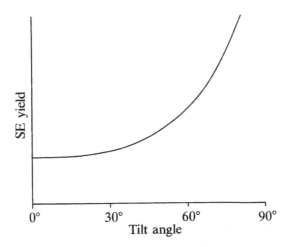

Fig. 4.2. The variation in secondary-electron yield with the angle of tilt of the specimen surface relative to the horizontal.

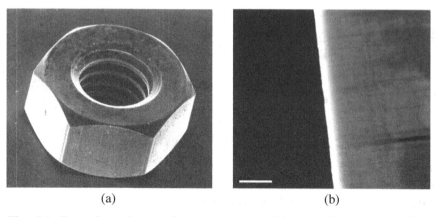

(a)   (b)

Fig. 4.3. Secondary-electron images, showing (a) three-dimensional effect resulting from the dependence of SE emission on the angle of the surface to the beam; and (b) edge effect (for a razor blade); scale bar = 5 μm.

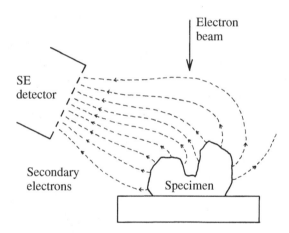

Fig. 4.4. Collection of secondary electrons from a three-dimensional specimen by a detector with a positively biassed grid.

secondary electrons, including those emitted from the far side of protruding features (Fig. 4.4) and from inside cavities, so that the image is relatively free of shadows. The detector should ideally be positioned adjacent to the top of the scanned raster (i.e. at the back of the specimen chamber), giving a 'top-lighting' effect.

More secondary electrons can escape from the sample at an edge (Fig. 4.5), causing it to appear bright (see Fig. 4.3(b)). This 'edge effect' is most prominent when a high accelerating voltage is used, owing to the greater penetration of the electrons.

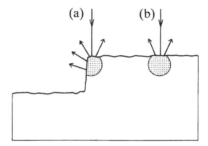

Fig. 4.5. The edge effect in a secondary-electron image: the signal is enhanced when the beam is (a) close to the edge, compared with (b) away from the edge.

Secondary-electron emission is not strongly composition-dependent and with coated samples emission is in any case governed mainly by the properties of the coating. The SE signal is, however, affected by variations in atomic number to some extent, mainly due to the production of secondary electrons by backscattered electrons (the yield of which is strongly Z-dependent) striking the polepiece of the electron lens, etc. (This effect can be minimised by covering the polepiece with material of low atomic number.) Also, the E–T detector output includes a contribution from direct detection of backscattered electrons, though this is relatively insignificant owing to the small solid angle subtended by the scintillator.

Examples of SE images of geological samples are given in Figs. 4.6–4.11.

4.4.2 *Topographic contrast in BSE images*

Owing to their relatively high energy, the paths of backscattered electrons are relatively unaffected by the positive bias on the E–T detector, and therefore travel in almost straight lines. With a detector located to one side of the specimen, a strong shadowing effect therefore occurs, since backscattered electrons emitted from the side of a 'hill' facing away from the detector are not detected. As well as this shadowing effect, the BSE yield exhibits a dependence on the angle between beam and specimen surface similar to that of the SE yield (Fig. 4.2), which contributes to the topographic contrast. Large-area scintillators and annular solid-state detectors (described in Section 3.11.2) are relatively non-directional but, with an annular detector divided into sectors, topographic contrast can be emphasised and compositional contrast suppressed by using the difference in the signals from opposite sectors (Figs. 4.12 and 4.13). Even in 'compositional' mode there is a topographic effect owing to variations in surface angle, hence these two effects are combined in cases such as that illustrated in Fig. 4.14.

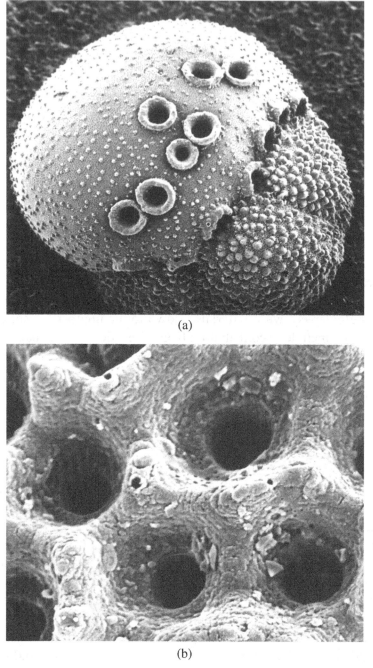

(a)

(b)

Fig. 4.6. Secondary-electron images of Globigerina: (a) 200×; and (b) 2000×.
(By courtesy of P. Pearson.)

Fig. 4.7. Secondary-electron image of aragonite in modern sediment (300 μm × 200 μm). (By courtesy of J. A. D. Dickson.)

Fig. 4.8. Secondary-electron image of diatoms from modern lake sediment (60 μm × 45 μm). (By courtesy of N. Cayzer and R. Thompson.)

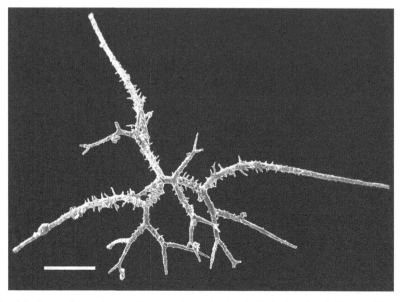

Fig. 4.9. Secondary-electron image of a spicular radiolarian (scale bar = 50 μm). (By courtesy of E. W. Macdonald.)

Fig. 4.10. Secondary-electron image (75 μm × 50 μm) of part of a filter-feeding Cambrian arthropod. (By courtesy of N. J. Butterfield.)

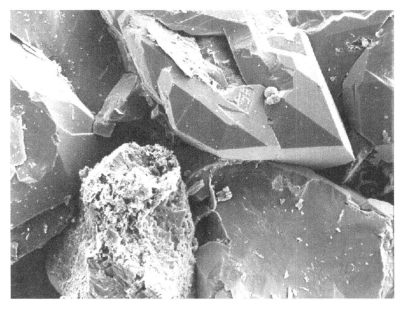

Fig. 4.11. Secondary-electron image of sandstone showing quartz grains, with corroded feldspar (bottom left). (By courtesy of S. Haszeldine.)

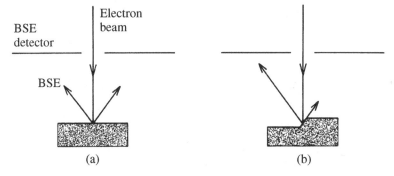

Fig. 4.12. Backscattered-electron signals recorded by opposite sectors of the detector are equal for a flat specimen (a) but differ where changes in topography occur (b); the difference signal therefore gives a topographic image in which the effect of composition is minimised, whereas the sum image emphasises compositional differences and minimises the effect of topography.

4.4.3 Spatial resolution

Secondary electrons have insufficient energy to travel distances greater than about 10 nm in solid materials, therefore only those produced close to the surface escape, and SE emission stimulated directly by the incident beam

(a)   (b)

Fig. 4.13. Backscattered-electron images of a copper grid mounted on an aluminium stub: (a) compositional mode, using omnidirectional BSE signal (brightness depends mainly on atomic number); and (b) topographic mode, using the difference between BSE signals from opposite segments of the detector (shows a topographic shadow effect, but no difference in mean brightness of copper and aluminium).

Fig. 4.14. A BSE image of mineral grains, showing combined compositional and topographic contrast (lighter grains – higher atomic number).

occurs only at or very near the point of entry. The spatial resolution in an image formed predominantly by such electrons is therefore dependent mainly on the beam diameter, though it is obviously limited to one pixel in a digital image, or the distance between lines in an analogue image.

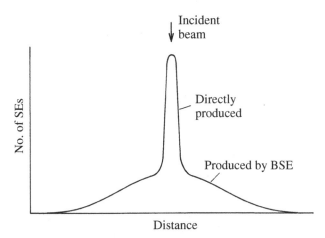

Fig. 4.15. The spatial distribution of secondary electrons: those produced directly by incident electrons entering the specimen originate from a small area close to the point of impact, whereas those produced by backscattered electrons as they leave originate from a large area.

To achieve high spatial resolution, the beam diameter must be made appropriately small by suitable choice of operating parameters (see Section 3.4). However, reducing the beam diameter entails unavoidable loss of current, as shown in Fig. 3.8 (for most purposes the practical lower limit on the beam current for SE images is a few picoamps). With a conventional tungsten electron source, the corresponding beam diameter is typically about 8 nm. With LaB_6 and FE sources (Section 3.2.1), minimum diameters of approximately 4 and 2 nm, respectively, can be obtained.

To attain the manufacturer's specified figure for resolution requires stray electromagnetic field and vibration to be below certain limits (which may be difficult to realise in some environments). Furthermore, the specification usually assumes a short working distance and an accelerating voltage of at least 15 kV (resolution tends to deteriorate at lower voltages).

The discussion so far has been concerned solely with secondary electrons produced by incident electrons as they enter the specimen. They are, however, also produced by backscattered electrons (Fig. 2.5(b)), and for high-atomic-number samples more are produced this way than directly by the incident beam. These BSE-related secondaries may emerge at a significant distance from the point of impact of the beam (Fig. 4.15). The fine detail reproduced by the directly generated secondaries is therefore superimposed on the relatively unsharp image formed by BSE-related secondaries (the effect of which can be reduced by using a low accelerating voltage, which decreases the electron range). However, at high magnifications BSE-related secondaries

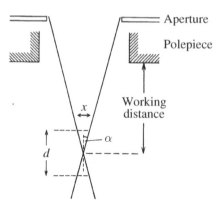

Fig. 4.16. Depth of focus: the width of the beam (x) resulting from its convergence angle has a negligible effect on image sharpness over a depth d (see text).

merely contribute a diffuse background. This effect is aided by using a high accelerating voltage, owing to the large range of the incident electrons (see Fig. 2.1).

4.4.4 Depth of focus

The depth of focus (or field) is the range of vertical distance over which image sharpness is not significantly degraded. The SEM has a much greater depth of focus than the optical microscope, owing to the small divergence (α) of the electron beam, which depends on the size of the final lens aperture and the working distance (Fig. 4.16). The depth of focus is given by $d = x/\alpha$, where x is the maximum acceptable beam broadening. Where maximum depth is required, it is desirable to minimise α by using a small-diameter aperture and a long working distance.

In a digital image, blurring by less than the width of one pixel has no visible effect. For a low-magnification image, with a raster approximately 1 mm in size, the pixel size is typically of the order of 1 μm: on putting $x = 1$ μm and assuming that $\alpha = 10^{-3}$ radians, we obtain $d = 1$ mm. A depth of focus approximately equal to the width of the scanned raster can thus be achieved.

4.4.5 Stereoscopic images

A stereoscopic effect can be produced by recording two images of the same area with the specimen tilt angle differing by a few degrees. With the detector in its usual position behind the specimen, the images must be rotated through 90° (anticlockwise), the low-angle image being on the left and the higher-angle one

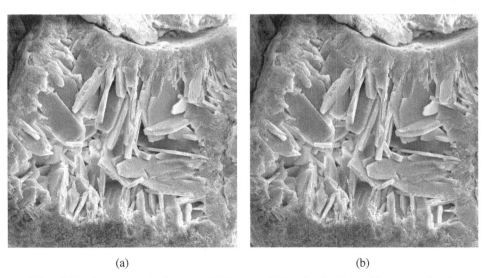

(a) (b)

Fig. 4.17. Stereoscopic images (400 μm × 400 μm) of clinoptilite crystals. A change in the tilt of the specimen causes small differences in the relative positions of features differing in height. See Plate 1 for a coloured anaglyph version.

on the right. Any shift occurring on tilting the specimen should be corrected if necessary.

A three-dimensional effect can be obtained by placing pairs of images such as those in Fig. 4.17 in a stereo viewer. Alternatively, by assigning contrasting colours (red and green or blue) to the images and combining them into one 'anaglyph', a stereoscopic effect can be obtained with the aid of coloured spectacles (Plate 1). Also, it is possible to produce 'live' stereo images using an arrangement whereby the angle of incidence of the beam is changed between successive scans and the alternating images are viewed through special spectacles that act as synchronised shutters.

Quantitative data on height variations of the sample can be derived from measurements of the displacements of recognisable features (Boyde, 1979; Goldstein *et al.*, 2003).

4.4.6 Environmental SEM

As described in Section 3.10.2, a 'low-vacuum' or 'environmental' SEM can be operated with relatively poor vacuum in the specimen chamber. This enables specimens containing volatiles such as water or oil to be examined (Uwins, Baker and Mackinnon, 1993), as well as fragile specimens and those on vacuum-incompatible mounts from which they cannot be removed (see Fig. 4.18). Also, uncoated specimens can be used without encountering charging effects.

Scanning electron microscopy ·

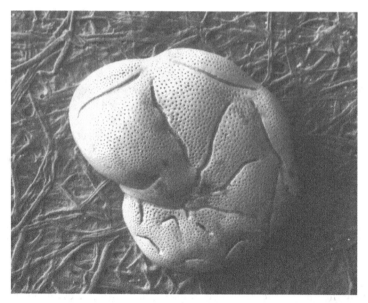

Fig. 4.18. Foraminifera on a cardboard mount, observed using low-vacuum SEM (1.0 mm × 0.85 mm). (By courtesy of P. D. Taylor.)

This is useful for delicate specimens that can be adversely affected by coating, for large specimens that are difficult to coat, and when coating is inadmissible on curatorial grounds. However, the resolution is inferior to that obtainable under high-vacuum conditions, owing to electron scattering.

4.5 Compositional images

The fraction of the electrons in the beam that are backscattered is strongly dependent on atomic number, Z (Fig. 2.4). The output of a BSE detector reflects this dependence in somewhat modified form because the detector is energy-sensitive and for high atomic numbers there is a larger proportion of high-energy electrons. Brightness in a BSE image of a specimen containing various phases of different compositions is thus a function of the mean atomic number $\bar{Z}$, which may be calculated to a first approximation using the mass concentrations of the elements present.

In principle a form of microprobe analysis based on measuring the BSE signal is possible. However, this requires prior knowledge of the identity of the elements present and is limited to binary (or quasi-binary) compounds, for example plagioclase feldspars in which the relatively small K content is linearly correlated with Ca (Ginibre, Kronz and Wörner, 2002). Fine-scale oscillatory zoning is resolved better than by X-ray analysis.

For compositional imaging the specimen should be well polished so that interference from topographic effects is minimal. In view of the insensitivity of BSE detectors to low-energy electrons, it is desirable to use an accelerating voltage of at least 15 kV, but the deterioration in resolution with increasing accelerating voltage should be taken into consideration.

Examples of BSE images of polished sections can be seen in Figs. 4.19 and 4.20. The limited minimum magnification obtainable in SEM images can be overcome by creating a 'montage' of individual images in order to cover a large area, for example a whole section (Fig. 4.21). The topic of BSE imaging in a geological context has been reviewed by Lloyd (1987) and Krinsley *et al.* (1998).

4.5.1 *Atomic-number discrimination in BSE images*

Values of $\bar{Z}$ for common minerals are listed in Tables 4.1 and 4.2. In many cases, of course, a given mineral has a range of compositions and therefore cannot be assigned a unique $\bar{Z}$ value: this applies particularly to ferromagnesian silicates, in which $\bar{Z}$ varies markedly with the Fe/Mg ratio. Low-Fe silicates have values clustering in the 10–11 range, owing to the strong influence of O ($Z = 8$) and Si ($Z = 14$). Silicates containing major amounts of K, Ca and Fe have intermediate values. For Fe–Ti oxides and most sulphides, $\bar{Z}$ is higher than for nearly all silicates and thus such phases show up clearly in BSE images of common types of rock. Discrimination between different sulphates and carbonates is usually fairly positive (though note that calcite and aragonite, for example, cannot be distinguished since they are chemically identical).

Although very often it is easy to distinguish between coexisting minerals in BSE images, there are cases (especially amongst silicates) in which $\bar{Z}$ values are so similar that this is difficult. Figure 4.22 shows the difference in atomic number (ΔZ) corresponding to a difference of 1% in the BSE coefficient. For silicates with $\bar{Z}$ of about 10, ΔZ as defined here is approximately 0.1. The choice of 1% is somewhat arbitrary and in practice the minimum detectable ΔZ depends on a number of factors. Even with expanded contrast, the ability to discriminate between areas of slightly different brightness is limited by noise (Section 4.6.1), which depends on detector efficiency, beam current and image-acquisition time. In the case of a 'live' image, the acquisition time is fixed and short, so the only way to reduce noise and improve atomic-number discrimination is to increase the beam current. A similar result can be achieved by using a stored image, with a longer acquisition time. The larger the areas concerned in the image, the smaller the minimum discernible difference in brightness (see Section 6.6), so high magnification is advantageous.

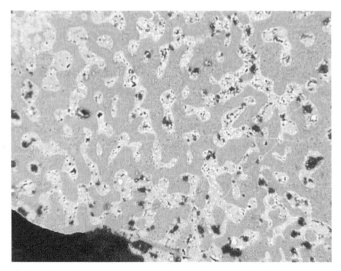

Fig. 4.19. BSE image (250 μm × 200 μm) of an echinoid ossicle; light areas –
dolomite, dark areas – calcite. (By courtesy of J. A. D. Dickson.)

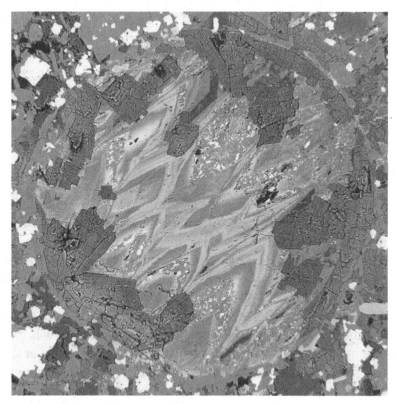

Fig 4.20. BSE image of a spherule (1 mm in diameter) in carbonate-rich lava,
showing Ba variations in zoned calcite.

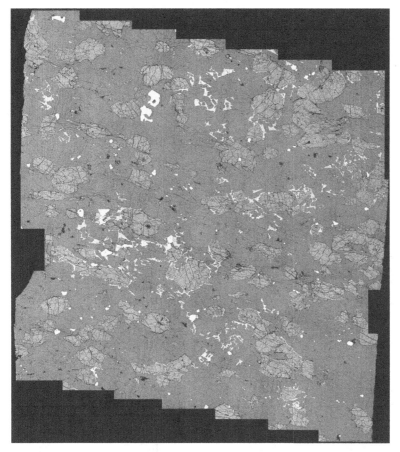

Fig. 4.21. A BSE image of a whole rock section (20 mm × 20 mm) comprised of a montage of individual smaller images. (By courtesy of J. Barreau.)

4.5.2 *Spatial resolution in BSE images*

Spatial resolution in BSE images is considerably worse than in SE images because many backscattered electrons travel significant distances within the sample before emerging. The width of the effective source area is typically about half the total range of the incident electrons in the sample and varies approximately as $E_0^{1.7}$, where E_0 is the electron energy. The best spatial resolution is therefore obtained with a low accelerating voltage, subject to limitations set by the sensitivity of the BSE detector for low energies.

Backscattered electrons that emerge some distance from the point of impact of the beam lose appreciable energy travelling through the specimen. On the other hand, electrons leaving the specimen close to the point of entry having suffered a large-angle deflection retain most of their initial energy. By

Table 4.1. *Mean atomic numbers of minerals, in alphabetical order*

Mineral	Mean Z	Mineral	Mean Z
Albite	10.7	Ilmenite	19.0
Almandine	15.6	Jadeite	10.7
Analcite	10.5	Kaolinite	10.4
Andalusite	10.7	Kyanite	10.7
Andradite	15.8	Lepidolite	11.1
Anglesite	59.4	Leucite	12.1
Anhydrite	13.4	Magnesite	9.4
Anorthite	11.9	Magnetite	21.0
Apatite	14.1	Malachite	18.6
Apophyllite	24.2	Monazite	38.7
Aragonite	12.4	Montecellite	13.8
Arfvedsonite	14.2	Mullite	11.1
Argentite	43.0	Muscovite	11.1
Arsenopyrite	27.3	Orthoclase	11.9
Baddeleyite	31.7	Pentlandite	22.9
Barite	37.3	Periclase	10.4
Benitoite	26.8	Perovskite	16.5
Bismuthinite	70.5	Platinum	78.0
Borax	7.7	Pyrite	20.7
Bornite	25.3	Pyrope	10.7
Brookite	16.4	Pyrrhotite	22.2
Calcite	12.4	Quartz	10.8
Cassiterite	41.1	Rhodochrosite	15.9
Celestine	23.7	Riebeckite	15.0
Celsian	27.2	Rutile	16.4
Cerussite	65.3	Serpentine	10.3
Chalcocite	26.4	Siderite	16.5
Chalcopyrite	23.5	Sillimanite	10.7
Chromite	19.9	Sodalite	11.1
Clinochlore	10.2	Spessartine	15.2
Cobaltite	27.6	Sphalerite	25.4
Columbite	24.6	Spinel	10.6
Copper	29.0	Spodumene	10.0
Corundum	10.7	Stibnite	41.1
Cuprite	26.7	Strontianite	25.6
Enstatite	10.7	Tantalite	65.4
Fayalite	18.7	Tephroite	17.2
Ferrosilite	16.9	Tetrahedrite	32.5
Fluorite	14.7	Titanite	14.7
Forsterite	10.6	Topaz	10.6
Galena	73.2	Ulvospinel	20.0
Glaucophane	10.6	Uraninite	82.0
Gold	79.0	Uvarovite	15.3
Graphite	6.0	Willemite	24.6
Grossular	12.9	Witherite	41.3
Gypsum	12.4	Wollastonite	13.6

Table 4.1. (*cont.*)

Mineral	Mean Z	Mineral	Mean Z
Haematite	20.6	Xenotime	24.2
Hercynite	15.3	Zircon	24.8
Humite	10.4	Zoisite	10.0

Table 4.2. *Minerals in order of mean atomic number*

Mean Z	Mineral	Mean Z	Mineral
6.0	Graphite	16.4	Brookite
7.7	Borax	16.4	Rutile
9.4	Magnesite	16.5	Perovskite
10.0	Zoisite	16.5	Siderite
10.0	Spodumene	16.9	Ferrosilite
10.2	Clinochlore	17.2	Tephroite
10.3	Serpentine	18.6	Malachite
10.4	Humite	18.7	Fayalite
10.4	Kaolinite	19.0	Ilmenite
10.4	Periclase	19.9	Chromite
10.5	Analcite	20.0	Ulvospinel
10.6	Forsterite	20.6	Haematite
10.6	Glaucophane	20.7	Pyrite
10.6	Spinel	21.0	Magnetite
10.6	Topaz	22.2	Pyrrhotite
10.7	Albite	22.9	Pentlandite
10.7	Andalusite	23.5	Chalcopyrite
10.7	Corundum	23.7	Celestine
10.7	Enstatite	24.2	Apophyllite
10.7	Jadeite	24.2	Xenotime
10.7	Kyanite	24.6	Columbite
10.7	Pyrope	24.6	Willemite
10.7	Sillimanite	24.8	Zircon
10.8	Quartz	25.3	Bornite
11.1	Lepidolite	25.4	Sphalerite
11.1	Mullite	25.6	Strontianite
11.1	Muscovite	26.4	Chalcocite
11.1	Sodalite	26.7	Cuprite
11.9	Anorthite	26.8	Benitoite
11.9	Orthoclase	27.2	Celsian
12.1	Leucite	27.3	Arsenopyrite
12.4	Aragonite	27.6	Cobaltite
12.4	Calcite	29.0	Copper
12.4	Gypsum	31.7	Baddeleyite
12.9	Grossular	32.5	Tetrahedrite
13.4	Anhydrite	37.3	Barite

Table 4.2. (*cont.*)

Mean Z	Mineral	Mean Z	Mineral
13.6	Wollastonite	38.7	Monazite
13.8	Montecellite	41.1	Cassiterite
14.1	Apatite	41.1	Stibnite
14.2	Arfvedsonite	41.3	Witherite
14.7	Fluorite	43.0	Argentite
14.7	Titanite	59.4	Anglesite
15.0	Riebeckite	65.3	Cerussite
15.2	Spessartine	65.4	Tantalite
15.3	Hercynite	70.5	Bismuthinite
15.3	Uvarovite	73.2	Galena
15.6	Almandine	78.0	Platinum
15.8	Andradite	79.0	Gold
15.9	Rhodochrosite	82.0	Uraninite

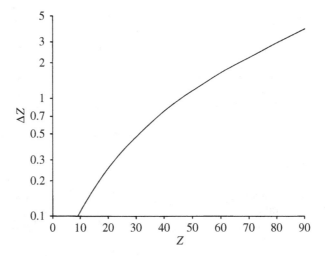

Fig. 4.22. Atomic-number difference (ΔZ) just distinguishable in a BSE image, as a function of atomic number (Z).

arranging to detect only these 'low-loss' electrons, considerably enhanced spatial resolution can be obtained.

4.5.3 The application of etching

By etching the specimen, compositional differences can be converted into topography and higher spatial resolution is then obtainable using SE imaging than is possible with backscattered electrons. This technique has been applied

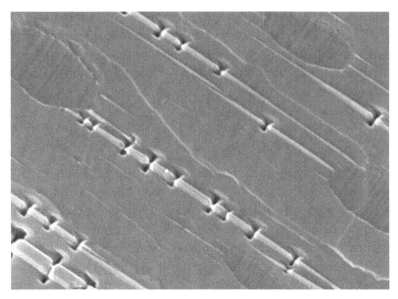

Fig. 4.23. Secondary-electron image of perthitic alkali feldspar (12 μm × 9 μm); the specimen was etched in HF vapour; two stages of exsolution can be seen: (1) rounded oligoclase lamellae with albite twinning; and (2) narrow albite lamellae with etch pits, showing the presence of dislocations between albite and K-feldspar host. (By courtesy of N. Cayzer.)

to exsolution lamellae in pyroxenes (Chapman and Meagher, 1975), for example, and to micro-perthite textures (Waldron, Lee and Parsons (1994); also, see Fig. 4.23). Etching methods are described in Section 9.4.

4.6 Image defects

Apart from inherent limitations as regards resolution, depth of focus, etc., SEM images may also be degraded in various ways, as described below.

4.6.1 Statistical noise

For images recorded at a fast scan rate, the number of electrons per pixel is small and is therefore subject to large statistical fluctuations, giving rise to 'noise'. Increasing the frame-scan time reduces noise, and using a long-persistence CRT screen or digital frame-store enables the whole of such an image to be viewed at once. However, the response to moving the specimen, adjusting the focus, etc. becomes slow. With a digital system the dwell time per pixel can be varied over a wide range; also noise can be reduced by post-acquisition smoothing (Section 4.7.1).

For photographic recording a single scan is used. An exposure time of the order of 1 min is usually sufficient to yield adequately noise-free images. Alternatively, a similar result can be obtained with a computer-based image store, with the advantage that it is possible to determine immediately whether the result is satisfactory, without the delay entailed in film processing.

4.6.2 Specimen charging

Owing to their low energy, secondary electrons are easily deflected by any charge present on the surface of the specimen. Insulating samples can be coated with a conducting material to prevent charging (Section 9.5), but, if the coating is defective, or there is dirt on the surface, charging artefacts will appear in the SE image (BSE images are affected less). If the whole specimen is charged, the image will be unstable. A completely non-conducting sample may cause an image of the interior of the specimen chamber to be produced by electrons repelled by the charged-up specimen. The remedy is to make sure that the coating is adequately conducting and connected to the holder, and the holder to earth. Charging can also be reduced by tilting the specimen, which increases the number of secondary and backscattered electrons.

The necessity for coating can be avoided by making use of the fact that the SE yield increases with decreasing incident electron energy so that, if a sufficiently low accelerating voltage is used (e.g. about 1 kV), the secondary and backscattered electrons leaving the specimen balance those arriving in the beam. Insulating specimens can thus be imaged without coating. Alternatively, charging can be avoided by using a relatively high pressure (Section 3.10.2).

4.6.3 Stray field and vibration

Stray alternating magnetic fields that are not fully suppressed by the screening of the column can cause straight edges in a scanning image to present a ragged appearance (Fig. 4.24). If it is impracticable to remove the source of the field, or install a field-cancelling system, the effect can be minimised by using a high accelerating voltage and minimum working distance.

Vibration can produce similar effects. It may be caused by mechanical vacuum pumps, for example, or by external factors such as traffic. It is usual for the column to be on anti-vibration mountings to minimise this effect.

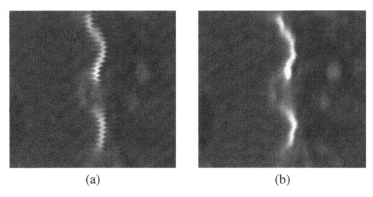

(a)   (b)

Fig. 4.24. The effect of stray mains-frequency electromagnetic field at high magnification (a); corrected by synchronising the line-scan rate (b).

4.6.4 Astigmatism

Figure 4.25 shows the effect of astigmatism, which, if not too serious, can be cured by adjustment of the stigmator (Section 3.3.1). Some instruments incorporate automatic astigmatism correction, which requires a test sample containing small spherical objects. If full correction is not attainable, it is necessary to clean the column apertures.

4.6.5 Coating artefacts

Coatings used to prevent specimen charging are not entirely structure-free and can cause artefacts in high-magnification SE images. These can be minimised by suitable choice of coating material (Section 9.5). Alternatively, the necessity for coating can be obviated by using a low accelerating voltage or a high specimen chamber pressure (see Section 3.10.2).

4.7 Image enhancement

Analogue display systems have limited provision for modifying the image. Contrast can be enhanced by increasing the amplifier gain and turning down the black level control (Fig. 4.26). Various effects can be obtained by varying the 'gamma' function, which is represented by the expression $I_{out} = I_{in}^{-\gamma}$. With γ greater than 1, dark parts of the image are lightened relative to lighter areas (Fig. 4.27), whereas a value of less than 1 results in the reverse effect. Another possibility is to apply differentiation to the electron signal, which has the effect of emphasising boundaries where the brightness suddenly changes (Fig. 4.28).

(a)   (b)

Fig. 4.25. Scanning image (a) with severe astigmatism; and (b) with astigmatism corrected by adjusting the stigmator.

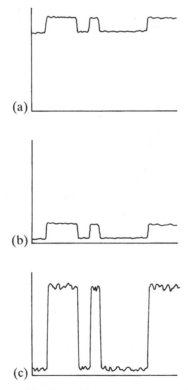

Fig. 4.26. Line-scan profiles in which low contrast in the original image (a) is enhanced by shifting the black level (b) and increasing the amplification (c).

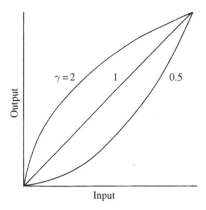

Fig. 4.27. 'Gamma function' used to modify image contrast.

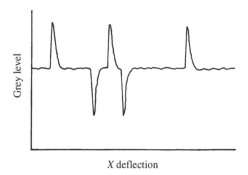

Fig. 4.28. A derivative signal obtained by differentiating the original signal in Fig. 5.8, to emphasise boundaries between areas of uniform brightness.

4.7.1 Digital image processing

One of the main advantages of digital SEM images is that they can be modified in many different ways after being recorded. Usually some image processing functions are available in the software provided with the instrument, but general-purpose 'off-line' software (as used for digital photographs) can also be employed. For this purpose images can be 'exported' in one of the commonly used formats, preferably uncompressed TIFF, which retains all the original information content, in preference to a compressed format such as JPEG.

The simplest way of representing an image is to make the brightness, or 'grey level', directly proportional to the recorded intensity of the relevant signal. Often it is advantageous to modify this relationship by applying an 'intensity transform' relating recorded intensity to grey level by a nonlinear function. In addition to the gamma function described in the preceding section, more complex functions may be used to manipulate image contrast in different

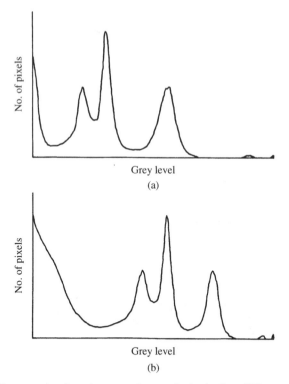

No. of pixels

Grey level

(a)

No. of pixels

Grey level

(b)

Fig. 4.29. Histograms showing numbers of pixels for different grey levels: (a) for the original image and (b) for the same image with expanded dark tones.

ways. A useful aid is the 'grey-level histogram', in which the numbers of pixels for each grey level are plotted: the distribution of pixels can be modified by applying an intensity transform, as illustrated in Fig. 4.29.

In digital images, differentiation (which in an analogue image affects only boundaries intersecting the horizontal lines) can be made equally effective for all boundary orientations, by means of a numerical operator such as a 'kernel', of the following form:

$$
\begin{array}{ccc}
-1 & -1 & -1 \\
-1 & 8 & -1 \\
-1 & -1 & -1
\end{array}
$$

For each pixel in turn the sum of the products of the adjacent pixel contents and the kernel coefficients is calculated. For areas of uniform brightness the result is zero, but a non-zero result is obtained on moving across a boundary, which enhances image sharpness (at the expense of increased noise).

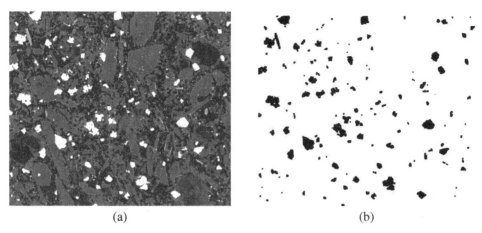

(a)   (b)

Fig. 4.30. (a) BSE image of a rock section (500 μm × 450 μm); (b) the same after applying 'threshold' and inverting, showing only grains of phases of high mean atomic number.

Noisy images can be improved by using a smoothing kernel, an example of which is:

$$\begin{array}{ccc} 1 & 1 & 1 \\ 1 & 8 & 1 \\ 1 & 1 & 1 \end{array}$$

This has the effect of averaging the intensity in neighbouring pixels, thereby reducing statistical variance (but also degrading the sharpness of the image). Too much smoothing causes undesirable artefacts to appear and it is preferable to obtain a less noisy image in the first instance, with increased beam current or acquisition time.

For some purposes it is useful to convert images into binary form in which each pixel is either black or white, which is done by applying a grey-level 'threshold'. For BSE images, in which grey level is related to atomic number, thresholding can be used to discriminate between phases of different compositions (Fig. 4.30), as a preliminary to modal analysis (Section 6.8).

4.7.2 *False colours*

The visual impact of an image is enhanced if the grey scale is replaced by a range of colours, represented by a 'look-up table' (LUT) relating colour to grey level. Such 'false-colour' images are used mostly for BSE or X-ray images, for which the signal used has a direct relationship to a quantifiable characteristic such as mean atomic number or elemental concentration, rather than for topographic images, for which the use of colour is less meaningful. The best effects are obtained by using a smooth colour scale, usually ranging from dark and 'cool' colours to light and 'hot' ones. An example of a false-colour image is shown in Plate 2.

4.8 Other types of image

Most SEM work involves the use of SE or BSE images showing predominantly topographic and compositional contrast, respectively. Various other types of image can be produced, however, as described in the following sections.

4.8.1 Absorbed-current images

The current flowing from the specimen to earth (equal to the incident beam current minus the current lost owing to backscattering and secondary-electron emission) can be amplified and used to produce an 'absorbed-current' image. Contrast in such images is reversed compared with normal images, since regions from which a large number of electrons are emitted appear dark rather than light.

Absorbed current is governed solely by the *number* of electrons leaving each point in the image (whereas in other imaging modes direction and energy influence detection efficiency), therefore shadow effects observed in SE and BSE images obtained with directional detectors are absent. Topographic contrast originates from variations in the local angle of the surface (which affects both forms of electron emission). Compositional differences also influence the image through the contribution of backscattered electrons.

4.8.2 Magnetic-contrast images

Magnetic domains can be revealed in SEM images by exploiting variations in the detection efficiency of secondary electrons resulting from deflection by the field immediately above the surface of the specimen. The biassed grid of the E–T detector attracts secondary electrons too strongly, giving only rather poor magnetic contrast, but this can be improved by placing an aperture in front of the detector to make it more selective. For further details, see Newbury *et al.* (1986).

An alternative way of viewing magnetic domains is the 'Bitter method', whereby a fluid containing suspended colloidal magnetite particles is applied to the surface and dried; the particles decorate domain walls and can be seen easily in a SE image (Moskowitz, Halgedahl and Lawson, 1988).

4.8.3 Electron backscatter diffraction images

The EBSD technique is increasingly commonly being used as an alternative to electron diffraction in the TEM, having the advantages of faster data acquisition and simpler specimen preparation. The crystallographic information obtained can be used for phase identification and texture analysis. A typical 'Kikuchi' pattern produced by EBSD is shown in Fig. 4.31. The boundaries of the bands represent positive and negative Bragg angles (typically of the order of $1°$) for a given set of crystallographic planes. The pattern is produced by electrons backscattered from close to the point of impact of the beam with very little energy loss (those that have lost more energy merely contribute to the diffuse background). Image processing can be applied to EBSD patterns to remove background, reduce noise and enhance contrast. Software for automatic indexing of the patterns and deriving data on lattice parameters and orientation is applicable to most mineral structures, with some exceptions. See Schwartz, Kumar and Adams (2000) for further details.

The signal from a 'forescatter detector' (Section 3.12.3) that detects electrons scattered over a limited angular range in the forward direction combines the effects of atomic number and orientation, and can be used to produce 'orientation-contrast' (OC) images (Prior *et al.*, 1996, 1999) in which different grey levels (or false colours) are related to orientation.

'Orientation maps' can be generated from patterns recorded at each point in a grid, which are obtained by moving the specimen. Orientation can be displayed by means of a colour code and boundary angles can also be displayed (Fig. 4.32, Plate 3). The spatial resolution is typically better than $1\,\mu m$ ($100\,nm$ is possible in favourable cases using a field-emission electron gun).

Normal specimen preparation results in a damaged surface layer, which seriously hinders the production of EBSD images, but, by using appropriate methods, a sufficiently damage-free surface can be obtained (see Section 9.3). Specimens should preferably be uncoated (approximate charge balance is obtained at the oblique incidence angle used for EBSD images), or coated with a very thin carbon layer, though this tends to reduce image contrast. Carbon contamination caused by the beam can seriously degrade the quality of EBSD patterns and should be minimised (see Section 3.10.1).

4.8.4 Cathodoluminescence images

Cathodoluminescence occurs in a range of different minerals and is caused by either crystal-structure defects or trace elements (see Section 2.8). Despite the fact that the origins of varying CL intensity and colour are often obscure, CL

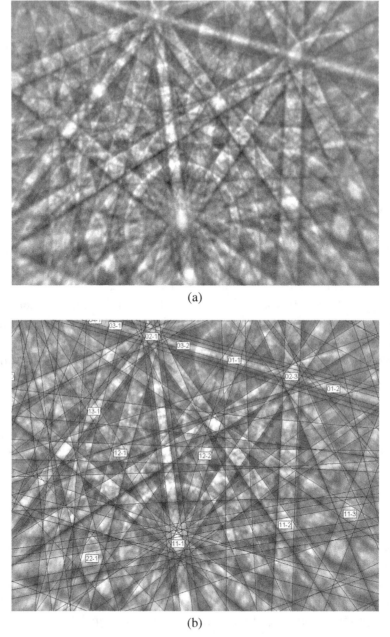

(a)

(b)

Fig. 4.31. EBSD pattern of zircon: (a) as recorded; (b) with indexing.
(By courtesy of P. Trimby.)

Fig. 4.32. Orientation map of calcite grains in marble derived from EBSD patterns. (By courtesy of G. Lloyd.) See Plate 3 for colour version.

images can play a useful petrographic role alongside BSE images and optical microscopy.

Quartz gives rather low CL intensity, but since neither BSE images nor X-ray maps generally provide useful information, SEM-CL has a particularly significant role. The causes of variations in CL emission are thought to be a combination of crystal defects and substitution of trace elements such as Al and Ti. In sandstones the difference between detrital and authigenic quartz (difficult to see in the optical microscope) shows clearly, the former being brighter, owing to crystallisation at a higher temperature giving a higher density of defects. Details of cementation, recrystallisation, fracture healing, etc. are also revealed. These effects are usually visible in panchromatic mode, but more information is obtainable when colour is taken into account (see Plate 4(a); also Laubach *et al.* (2004)). Demars *et al.* (1996) noted the existence of an emission band in the UV region, which is apparently related to coupled substitution of Al and Li, and occurs with greater intensity in authigenic quartz. Growth zones that are otherwise invisible can also be observed in CL images of quartz in volcanic rocks (Watt *et al.*, 1997) and granite (D'Lemos *et al.*, 1997), and filled microfractures appear bright in CL images, owing to their high density of defects (Watt, Oliver and Griffin, 2000). Haloes due to radiation damage can be seen in Plate 4(b). Deformation lamellae in quartz that has experienced meteorite-impact shock show as black lines, since the very high defect density suppresses CL emission (Boggs *et al.*, 2001).

There are greater compositional variations amongst feldspars, giving rise to a range of CL intensity and colour. Cathodoluminescence images reveal effects related to crystal growth and alteration that are somewhat similar to those of quartz. Differences between plagioclase and orthoclase provide a convenient means of identifying these minerals in fine-grained intergrowths.

Calcite is well known for its intense orange CL emission caused by Mn. Dolomite also luminesces, but with a redder colour, which is useful as an aid to identification. The marked banding observed in calcite overgrowths is controlled by varying concentrations of both Mn and Fe, the latter having a quenching effect. Though this is easily visible with a CL microscope, SEM-CL enables finer detail to be observed, especially in darker regions (Fig. 4.33). The long decay time of CL emission in some minerals can be minimised by increasing the dwell time or, in the case of calcite, by selecting a wavelength in the violet–UV region (Reed and Milliken, 2003).

Cathodoluminescence images of zircon commonly show considerable detail related to growth processes. Since the CL emission is at least partly dependent on compositional variations (typically involving rare earths and yttrium), BSE images usually show similar general features, though CL images often reveal finer detail. Cathodoluminescence images are used to characterise different groups of detrital zircons in sediments and for selecting areas for ion microprobe U–Pb dating (see Fig. 4.34).

For further information on applications of CL, see Marshall (1988) and Pagel *et al.* (2000).

4.8.5 Charge-contrast images

Secondary-electron images of uncoated samples of materials with poor electrical conductivity obtained with an 'environmental' or low-pressure SEM (Section 3.10.2) may, under certain conditions, exhibit contrast related to the local charge density in the surface region (Watt, Griffin and Kinny, 2000). To achieve this, the gas pressure in the sample chamber should be adjusted to a level at which specimen charging is only partly neutralised by positive ions.

Fig. 4.33. SEM-CL image of zoned calcite with a complex growth history (1 mm × 1 mm); finer detail can be resolved than with a CL microscope. (By courtesy of M. Lee; first published in *Microscopy and Analysis*, no. 79, September 2000, p.15.)

Variations in the local surface potential then influence secondary-electron emission, producing contrast dependent on conductivity. Figure 4.35 shows an example of such a 'charge-contrast image' (CCI). This type of contrast is related to crystal defects connected with either growth processes or trace elements and is correlated with contrast of similar origin observed in cathodo-luminescence images (see Section 4.8.4). Higher spatial resolution is attainable in CCI images and they can be obtained for a wider range of sample types, including phases that do not luminesce.

4.8.6 *Scanning Auger images*

The combination of an Auger electron spectrometer with an SEM to give a 'scanning Auger microscope' (SAM) is described in Section 3.12.1. Elemental

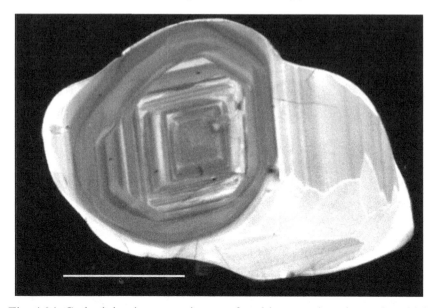

Fig. 4.34. Cathodoluminescence image of multi-generation zircon, in which brightness is related to composition and defect density, used as a guide for selection of areas for ion microprobe U–Pb dating. The original zoned core (~2850 Ma) has been overgrown by a weakly luminescent band (~2550–2510 Ma) and a highly luminescent banded and sector-zoned overgrowth (~2490 Ma). Scale bar = 100 μm. (By courtesy of N. Kelly.)

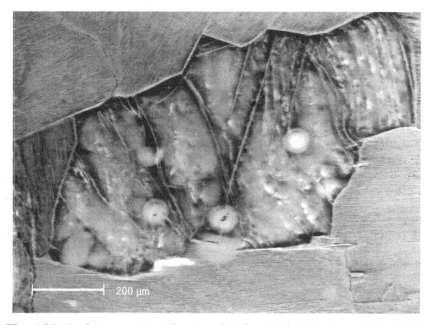

Fig. 4.35. A charge-contrast image of deformed biotite in granite (Watt, Griffin and Kinny, 2000), showing folded cleavage, linear fractures and radiation haloes caused by small monazite inclusions (for explanation of the contrast mechanism, see Section 4.8.5). (By courtesy of B. Griffin.)

distribution 'maps' are formed by the signal obtained with the spectrometer set to an Auger line of the element of interest. For this surface-sensitive technique the use of a conductive coating is inappropriate and specimen charging is therefore a problem in geological applications. It can be avoided, however, by using a low accelerating voltage and low beam current (e.g. 3 kV and a few nanoamps) and by tilting the specimen, which increases the number of back-scattered and secondary electrons leaving the sample.

The spatial resolution is very high with respect to depth, owing to the small 'escape depth' of Auger electrons. Lateral resolution is limited by the contribution of Auger electrons produced by backscattered electrons leaving the specimen, but is typically a fraction of 1 μm (considerably better than for X-ray images). The technique thus provides a method of surface analysis with high lateral resolution. Geological applications have been reviewed by Hochella, Harris and Turner (1986) and Hochella (1988).

5

X-ray spectrometers

5.1 Introduction

X-ray spectrometers are of two kinds. The energy-dispersive (ED) type records X-rays of all energies effectively simultaneously and produces an output in the form of a plot of intensity versus X-ray photon energy. The detector consists of one of several types of device producing output pulses proportional in height to the photon energy. The wavelength-dispersive (WD) type makes use of Bragg reflection by a crystal, and operates in 'serial' mode, the spectrometer being 'tuned' to only one wavelength at a time. Several crystals of different interplanar spacings are needed in order to cover the required wavelength range. Spectral resolution is better than for the ED type, but the latter is faster and more convenient to use. X-ray spectrometers attached to SEMs are usually of the ED type, though sometimes a single multi-crystal WD spectrometer is fitted. Electron microprobe instruments are fitted with up to five WD spectrometers, and often have an ED spectrometer as well. These two types of spectrometer are described in detail in Sections 5.2–5.4 below.

5.2 Energy-dispersive spectrometers

The modes of operation and characteristics of the detectors and associated electronics used as ED spectrometers are described in the following sections.

5.2.1 Solid-state X-ray detectors

In ED spectrometers the X-ray detection medium is a semiconductor (either silicon or germanium), in which the valence band is normally fully occupied by electrons. The valence and conduction bands are separated by an energy gap (1.1 eV for Si, 0.7 eV for Ge), and at room temperature very few electrons have sufficient thermal energy to jump this gap: the conductivity is therefore normally very low. When an X-ray photon is absorbed it generates Auger electrons and photo-electrons (Section 2.6), which dissipate their energy partly by raising electrons from valence to conduction band. The arrival of each photon thus creates a brief pulse of current caused by electrons in the latter and 'holes' in the former, moving in opposite directions under the influence of the bias voltage applied to the detector.

The mean energy used in generating one electron–hole pair is 3.8 eV for Si (2.9 eV for Ge). The size of the output pulse depends on the number of such pairs, which is given by the X-ray energy divided by the mean energy. Hence a 1.487-keV Al Kα photon, for example, produces an average of 391 electron–hole pairs in a Si detector, whereas a Ni Kα photon (7.477 keV) produces 1970 (the actual numbers are subject to some statistical fluctuation).

Even highly refined silicon contains impurities, which have undesirable effects. These are counteracted by introducing lithium using a process known as 'drifting' – hence the name 'lithium-drifted silicon', or 'Si(Li)', detector. Germanium detectors are usually made of high-purity material ('HPGe'), which does not require the addition of Li. A typical Si(Li) detector consists of a silicon slice about 3 mm thick, with an area of 10 mm^2 (though larger sizes are available). The front surface is covered by a thin layer of gold, which serves as a contact for the bias voltage. The rear is connected to a field-effect transistor (FET), which acts as a preamplifier. The detector and FET are mounted on a copper rod, the other end of which is immersed in liquid nitrogen, and the whole assembly is sealed inside an evacuated housing, or 'cryostat' (Fig. 5.1). (Mechanical refrigeration can be used instead to obviate the need for liquid nitrogen.) The detector must be protected from damage resulting from warming up while the bias voltage is on, by means of a temperature sensor that switches off the voltage.

X-rays reach the detector via a 'window' capable of withstanding atmospheric pressure, so the vacuum chamber of the instrument to which it is attached can be vented to air safely. A typical beryllium window about 8 μm thick absorbs X-rays of energy below about 1 keV, but low-energy X-rays can be detected if a thin polymer window is used instead. To provide sufficient strength to withstand atmospheric pressure, such windows are supported by a grid. Thin-window detectors are sensitive to light, and in instruments with a microscope the lamp must be switched off for X-ray spectrum acquisition.

An alternative type of detector is the 'silicon drift detector' ('SDD'), which is different in construction and does not involve Li compensation ('drift' in this context refers to the motion of the charge carriers). The main advantages are the ability to operate at count-rates above 10^5 counts s^{-1}, and the fact that moderate thermo-electric cooling is sufficient, obviating the need for liquid nitrogen.

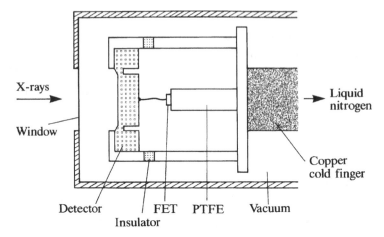

Fig. 5.1. Mounting arrangements for a solid-state detector used for energy-dispersive spectrometry.

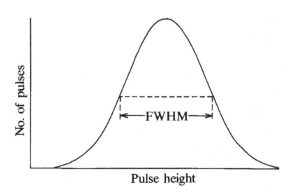

Fig. 5.2. Gaussian pulse-height distribution: FWHM = full width at half maximum.

5.2.2 Energy resolution

The number of electron–hole pairs as calculated in the previous section is a mean value. In reality the number is subject to statistical fluctuations, and X-rays of a given energy E produce a spread of pulse heights in the form of a Gaussian distribution (Fig. 5.2). The width is represented by the full width at half maximum ('FWHM'), which can be expressed in energy units as ΔE, given by

$$\Delta E^2 = kE + \Delta E_n^2, \qquad (5.1)$$

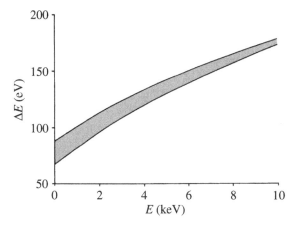

Fig. 5.3. The energy resolution of an ED spectrometer as a function of the X-ray energy (*E*): ΔE is the full width at half maximum of the peak (see Fig. 5.2); the shaded area represents the range of values obtained with Si(Li) detectors.

where the value of the constant *k* is 2.53 for Si (1.93 for Ge). The first term is determined by statistics and the second represents the effect of electronic noise in the detector and preamplifier. The latter varies somewhat between different detectors and also increases with increasing count-rate. The variation of ΔE with *E* takes the form shown in Fig. 5.3. Resolution is conventionally defined as the FWHM peak width of the Mn Kα line (5.89 keV): the lowest practical values (obtainable at low count-rates only) are about 128 eV for Si and 115 eV for Ge.

5.2.3 Detection efficiency

The X-ray collection efficiency of an ED spectrometer is determined primarily by the solid angle subtended by the detector, which is given by the area of the detector divided by the square of its distance from the source. A large solid angle is desirable for applications requiring a low beam current, in which case the detector should be as close as possible to the specimen. However, sometimes it may be necessary to reduce the efficiency: for example when using WD and ED spectrometers simultaneously, where the former requires a relatively high beam current. To avoid overloading the detector, the sensitivity can be reduced by moving the detector away from the specimen, when a means for doing this is available, or by introducing an aperture in front of the detector.

The efficiency of detection of X-rays reaching the detector is close to 100% over a wide energy range. Above about 20 keV the efficiency of a Si detector of normal thickness falls because only partial absorption within the detector

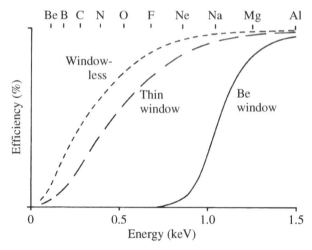

Fig. 5.4. ED detector efficiency in the low-energy region, with various entrance windows (schematic only; the exact shape of the curve depends on window thickness and composition).

occurs (Ge detectors are more efficient in this region owing to their greater absorbing power). Where a beryllium window is fitted, efficiency falls below about 2 keV, owing to absorption in the window, and is effectively zero below about 0.7 keV, whereas useful efficiency is maintained even at low energies when a thin window is fitted. Absorption also occurs in the gold contact layer on the surface of the detector and in the 'dead layer' of silicon between the gold layer and the active region. Figure 5.4 shows the detection efficiency as a function of energy with various windows.

Sometimes absorption is increased by a film of vacuum-pump oil condensed on the window (which is colder than its surroundings). The oil can be washed off (following the manufacturer's instructions) and condensation may be avoided by installing a low-power heater to raise the temperature of the detector housing slightly. This may also be used to remove ice formed from water vapour diffusing through the window. The symptom of either of these phenomena is reduced relative sensitivity to low-energy X-rays. The build-up of ice or oil can be monitored by measuring the intensity ratio of suitably chosen peaks (e.g. Cu Lα/Cu Kα) under standardised conditions.

5.2.4 Pulse processing and dead-time

The output pulses from the preamplifier are amplified to a size suitable for pulse-height analysis. To minimise the effect of noise, the signal is averaged over a time interval (typically a few tens of microseconds) defined by a

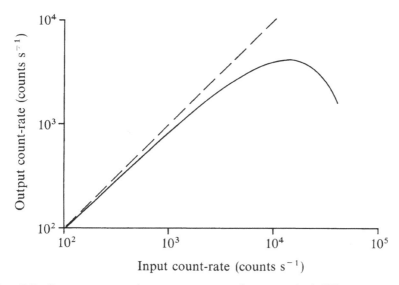

Fig. 5.5. Output versus input count-rate for a typical ED system: the maximum output is governed by the extendable dead-time of the pulse-processing electronics.

parameter known as the 'time constant', or 'process time'. While each pulse is being processed the system is 'dead' – that is it does not respond to any further pulses arriving from the detector. The time from the arrival of a pulse to the moment when the system is 'live' again is the 'system dead-time' (t), which is related to the time constant.

The dead-time of an ED system is 'extendable', meaning that if another pulse arrives during the processing of the preceding one, it extends the dead period by a further time t. For such a system the input count-rate n (the rate of arrival of pulses from the detector) and the output count-rate n' (the rate at which processed pulses are accumulated in the spectrum) are related thus:

$$n' = n \exp(-nt). \tag{5.2}$$

Equation (5.2) leads to the behaviour illustrated in Fig. 5.5: the output count-rate increases with increasing input count-rate at first, reaches a maximum, and then falls. At the maximum, $n = t^{-1}$ and $n' = (et)^{-1}$. A higher input rate is pointless since the rate at which counts are accumulated in the spectrum then falls below the maximum.

The 'per cent dead-time' represents the percentage of elapsed 'real' time for which the system is dead, which is equal to $100[1 - \exp(-nt)]$. Normally a warning is given if this exceeds 50%: this is partly because it is close to the

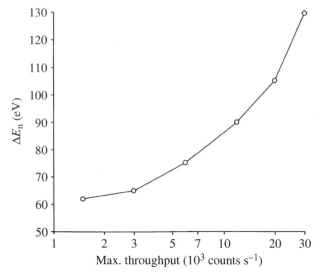

Fig. 5.6. The relationship between ΔE_n, the contribution of electronic noise to the energy resolution of the ED detector (see the text), and maximum 'throughput'.

point of maximum 'throughput' (at 68% dead-time), but also because undesirable behaviour is liable to occur at higher count-rates.

The loss of effective counting time can be compensated automatically by switching off the 'clock' controlling the counting time during the dead periods. The spectrum is recorded for a live time selected by the user, the real time which elapses during the recording period being extended to allow for the dead-time. For example, with 30% dead-time a real time of 130 s is required in order to accumulate a spectrum with a live time of 100 s.

The best energy resolution is obtained with a long time constant. As this is reduced, ΔE_n in Eq. (5.1) increases and the resolution worsens, but pulses can be recorded at a faster rate. The relationship between energy resolution and maximum recording rate (as determined by the dead-time) for a typical system is shown in Fig. 5.6. The energy resolution at a reasonably high count-rate is more important for most purposes than the best possible resolution, which is obtainable only at a low rate.

5.2.5 *Spectrum display*

The train of amplified pulses from the detector is converted into a spectrum by means of a 'multichannel pulse-height analyser', which measures the height of each incoming pulse and assigns it to a 'channel'. The recorded spectrum takes

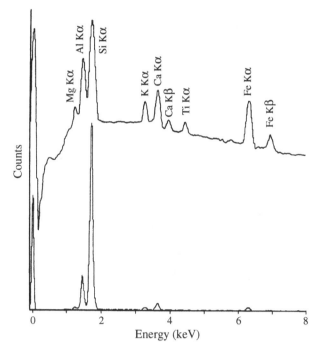

Fig. 5.7. The energy-dispersive X-ray spectrum of a silicate mineral, consisting of a histogram of counts per channel, using logarithmic and linear scales (upper and lower curves, respectively).

the form of an array of numbers representing the contents of each channel and is displayed as a histogram (Fig. 5.7). Most X-ray lines of interest fall within the range 0–10 keV, requiring 1000 channels of width 10 eV, or 2000 of width 5 eV. Correct energy calibration can be obtained by adjustment of zero and gain controls so that known X-ray lines (a minimum of one at low energy and one at high energy) appear at the correct points on the energy scale.

Various display facilities are provided: for example, the spectrum can be expanded on either axis to facilitate inspection of particular features. As well as the normal linear intensity scale, a logarithmic mode is an optional alternative: this has the advantage that large and small peaks are visible simultaneously. Markers showing the positions of elemental peaks can be inserted to assist identification. A useful feature is the capability to display two spectra simultaneously so that similarities and differences can be studied.

In addition to the spectrum itself, it is usual to have additional data displayed on the screen: typically these include current set-up information such as electron volts per channel, energy range, type of display (logarithmic or linear), intensity range, count-rate, preset live-time, elapsed time and per cent dead-time, together with information such as date, time and spectrum label.

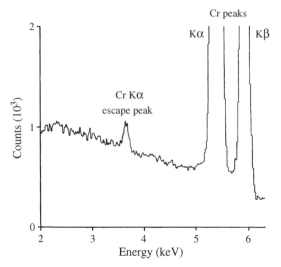

Fig. 5.8. The 'escape peak' in the ED spectrum of Cr, occurring 1.74 keV below the main 'parent' peak (see text for explanation).

5.2.6 *Artefacts in ED spectra*

There are certain artefacts in ED spectra of which the user should beware. One of these is the 'escape peak' – a small 'satellite' appearing 1.74 keV below its 'parent' peak in the case of a Si(Li) detector (Fig. 5.8), which is produced by the following mechanism. After absorption of an X-ray photon in the detector, a Si K photon may be emitted and, though usually this will be absorbed within the detector, there is a finite probability that it may escape. If it does, the output pulse height is reduced owing to the loss of the energy carried by the Si K photon. The probability of escape depends on the energy of the incident photon (which determines how deeply it penetrates the detector) but is generally less than 1%, so escape peaks are significant only when the parent peak is large. With a Ge detector escape peaks are not usually observable.

Another related phenomenon occurring with Si(Li) detectors is the appearance of a spurious Si K peak in the spectrum, giving the impression of the presence of a small amount of Si (a fraction of 1%) even in samples containing none. The size of the Si peak varies with the content of the spectrum.

Small artefact peaks may also appear at energies corresponding to the sums of the energies of major peaks in the spectrum (Fig. 5.9). These 'sum peaks' are caused by pairs of pulses arriving so close together in time that the pulse processor sees them as one. This effect is greatly reduced by electronic 'pile-up rejection', but can never be eliminated totally. The probability of such

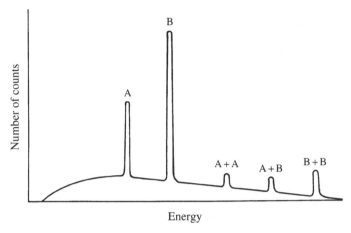

Fig. 5.9. 'Sum peaks' in the ED spectrum, occurring at energies equal to the sums of the energies of the main peaks.

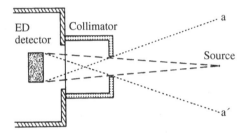

Fig. 5.10. Collimator attached to the front of an ED detector, limiting acceptance to X-rays originating between a and a'.

coincidence between pulses is a function of the count-rate: sum peaks are therefore significant only at high count-rates and, if troublesome, can be reduced by decreasing the count-rate.

ED detectors have a wide angle of acceptance and do not discriminate between X-rays produced at the point of impact of the electron beam and those generated elsewhere by stray electrons. The spectrum may thus contain an unwanted contribution from other regions of the specimen, the specimen holder, etc. This can be minimised by means of a collimator in front of the detector, which restricts the range of angles accepted (Fig. 5.10). However, it is impossible to achieve perfect discrimination, and small spurious peaks may still occur.

Some backscattered electrons may have enough energy to penetrate the detector window (especially if this is of the ultra-thin type) causing spurious output pulses that increase the background level. This effect (which is greatest with a high accelerating voltage) can be prevented by fitting an electron trap in the form of a permanent magnet mounted in front of the detector.

5.3 Wavelength-dispersive spectrometers

Wavelength-dispersive (WD) spectrometers are distinguished from the energy-dispersive (ED) type by the fact that the X-rays are 'dispersed' according to their wavelength by means of Bragg reflection. WD spectrometers give high spectral resolution but generally lower intensity for a given beam current than the ED type: the two are thus complementary, as discussed in more detail in Section 5.4.

5.3.1 Bragg reflection

In certain directions waves scattered from successive layers of atoms in a crystal are in phase and their intensity is enhanced. This is illustrated in Fig. 5.11, where the difference in path length between rays ABC and A'B'C' is an integral multiple of the wavelength (λ). This condition results in the reflection of X-rays of a given wavelength by atomic layers with interplanar spacing d at a certain angle of incidence and reflection known as the 'Bragg angle' (θ). The relationship between these variables is given by Bragg's law,

$$n\lambda = 2d \sin \theta, \tag{5.3}$$

which follows directly from the difference in path lengths between successive planes. The integer n is the 'order' of reflection. The most intense reflections, which are normally used in WD analysis, are those of the first order ($n = 1$). Higher orders add unwanted peaks to the spectrum, but their intensity is relatively low and they can be suppressed by pulse-height analysis (Section 5.3.4).

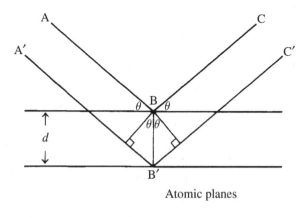

Fig. 5.11. Bragg reflection: diffracted rays are in phase when distance A' B' C' differs from ABC by an integral number of wavelengths.

Table 5.1. *Crystals used in WD spectrometers, with values of 2d*

Crystal	$2d$ (Å)
LiF	4.026
PET	8.742
TAP	25.9

Table 5.2. *Evaporated multilayers for long-wavelength WD spectrometry*

$2d$ (Å)	Components	K lines detected
60	W–Si	F, O
100	Ni–C	C, B
160	Mo–B_4C	Be, N

It follows from Eq. (5.3) that the wavelength range (for $n = 1$) is limited for a given value of $2d$, and several crystals of different spacings are therefore needed in order to cover the whole range of wavelengths of interest. The crystals normally used are listed in Table 5.1 and their wavelength coverage is shown in Fig. 5.12. Since the wavelength ranges overlap there is sometimes a choice between two crystals for a given X-ray line (for example, the Ca Kα line is reflected by LiF at 56.5° and by PET at 22.6°). Where a given wavelength is covered by two crystals, the one with the larger d and hence lower Bragg angle gives higher intensity but poorer resolution (see Fig. 5.13).

Crystals with spacings larger than that of TAP are not available, but artificial layered structures, or 'pseudo-crystals', enable the K lines of light elements down to Be (atomic number 4) to be covered. In lead stearate, layers of Pb atoms are separated by hydrocarbon chains, giving $2d = 100$ Å, and similar pseudo-crystals with larger spacings also exist. However, 'multilayers' produced by evaporating alternate layers of heavy and light elements (e.g. W and Si) with contrasting X-ray scattering powers give considerably higher intensities (but worse resolution) and have the additional advantage that reflections of order higher than 2 are very weak, reducing the likelihood of interference from lines of shorter wavelength. For light-element analysis it is preferable to use multilayers optimised for different wavelengths by suitable choice of elements and layer thicknesses (see Table 5.2).

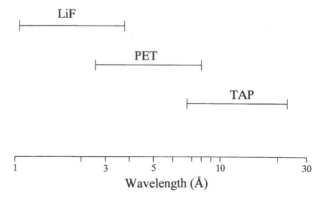

Fig. 5.12. Wavelength coverages of crystals used in WD spectrometers.

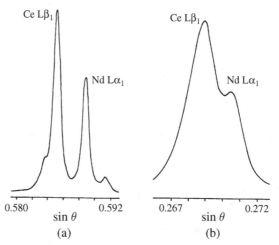

Fig. 5.13. The effect of choice of crystal on resolution in a WD spectrum: peaks are resolved with LiF (a) but not with PET (b).

5.3.2 Focussing geometry

Bragg reflection typically occurs over a range of angles of less than 0.01° and, with a point source of X-rays and a flat crystal, reflection of a particular wavelength occurs over only a small part of the crystal. A larger reflecting area can be obtained, however, if the crystal is curved. In the usual form of WD spectrometer, source, crystal and detector are located on an imaginary 'Rowland circle' (Fig. 5.14). The atomic planes are curved to twice the radius of this circle (r), which makes the Bragg angle the same at all points. Ideally the surface of the crystal should lie on the Rowland circle, requiring the surface to be ground to a radius r ('Johansson geometry'). It is relatively easy to grind

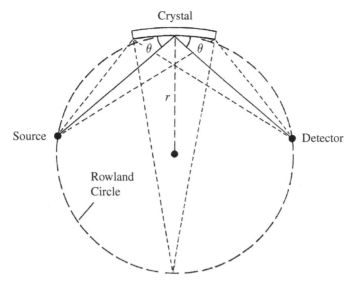

Fig. 5.14. Rowland circle geometry: a constant Bragg angle is obtained when the source, crystal and detector lie on the circumference of a circle.

LiF crystals, but PET and TAP are more difficult. 'Johann geometry', in which the crystal is curved to the appropriate radius but not ground, is easier to achieve but entails some sacrifice in resolution.

By placing a narrow slit in front of the counter, the peaks can be made sharper (at the expense of reduced intensity). The enhanced resolution is advantageous only in rare cases in which lines are so close that they are not normally resolved. Otherwise it is better not to use a narrow slit (especially for quantitative analysis), because it makes the peak intensity more sensitive to error in the Bragg-angle setting.

Crystals of larger than normal dimensions are sometimes available as an option, offering enhanced intensity, with some sacrifice in resolution. In many applications resolution is not critical, but it affects the peak-to-background ratio, which cancels out some of the benefit of the increased intensity, as far as trace-element detection is concerned (Section 8.5).

Defocussing effects

The focussing geometry of a WD spectrometer is correct only with the X-ray source in its normal position, that is, on the axis of the column and with the surface of the specimen in the correct plane. Movement of the source in the z direction causes a change in the Bragg angle (Fig. 5.15(a)), thereby shifting the position of the peak relative to the wavelength scale, which is particularly important in quantitative analysis. In an electron microprobe the

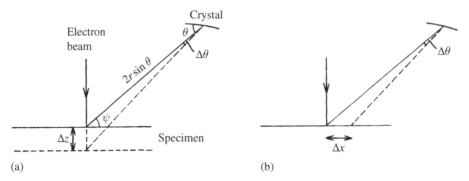

Fig. 5.15. Changes in Bragg angle ($\Delta\theta$) caused by displacement of the X-ray source from its correct position: (a) in the z direction (height); and (b) in the x direction (lateral shift).

specimen can be brought to the focus of the optical microscope by means of the stage z movement, thereby avoiding the defocussing effect. In SEMs the specimen plane is less well defined, the depth of focus in a scanning image being much greater. A similar effect occurs if the beam is moved in a radial direction relative to the Rowland circle (Fig. 5.15(b)). It is therefore desirable to move the specimen rather than the beam when selecting different points for analysis. Also, scanning images are liable to show intensity loss at the edges (see Section 6.4).

In Fig. 5.15 the spectrometer is assumed to be mounted vertically. The effect of specimen-height variation is greatly reduced if the spectrometer is mounted in an 'inclined' configuration (Fig. 5.16). This orientation is favoured when the specimen height is poorly constrained, as in the case of SEMs. It is less usual in electron microprobes, where it reduces the number of spectrometers that can be fitted.

5.3.3 Design of WD spectrometers

In practical spectrometer designs the centre of the Rowland circle (Fig. 5.14) is not in a fixed position. Instead the crystal moves along a linear track (aligned with the X-ray source), the correct Bragg angle (θ) being obtained by means of a mechanical linkage, which also moves the detector along the appropriate path. The source–crystal distance (x) is related to θ by the expression $x = 2r \sin \theta$, where r is the radius of the Rowland circle. Since $\sin \theta = \lambda/(2d)$ (for $n = 1$), it follows that $x = (r/d)\lambda$. The source–crystal distance is thus a linear function of wavelength and the required wavelength is obtained by moving the crystal along its track. The scale is calibrated by reference to a known X-ray line. WD

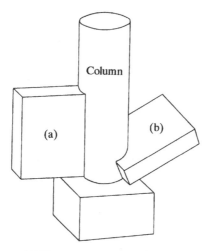

Fig. 5.16. Orientation of WD spectrometers: (a) vertical – occupies less space but is sensitive to specimen-height variations; and (b) inclined to the horizontal – insensitive to specimen height, but occupies more space, so fewer spectrometers can be fitted.

spectrometers are usually provided with two or more crystals mounted on a turret, in order to extend the wavelength range.

Normally WD spectrometers are evacuated to eliminate X-ray absorption by air, which, though not serious for X-rays at the short-wavelength end of the range of interest, is prohibitive at the long-wavelength end. High vacuum is not essential, and in some instruments the spectrometers are pumped by a rotary pump only, with the spectrometer separated from the column by a thin window.

Electron microprobe instruments are commonly fitted with up to five vertical WD spectrometers around the column. This has the advantage that crystal changes can be avoided and time saved in multi-element analyses. Usually SEMs can be fitted with only one spectrometer and therefore are less efficient for WD analysis.

An important parameter is the 'X-ray take-off angle', which is defined as the angle between the surface of the specimen and the X-ray path to the spectrometer. If this is too low, the absorption of emerging X-rays is excessive (Section 7.7.2). On the other hand, a high angle conflicts with other considerations, including the desirability of a short final lens working distance. A reasonable compromise is a value of about 40°, which is typical for contemporary instruments (though note that the effective take-off angle in SEMs is dependent on specimen tilt).

Alternatives to the conventional 'fully focussing' type of WD spectrometer described above have been developed specifically for SEMs. In the 'semi-focussing'

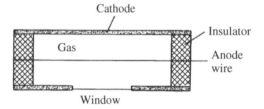

Fig. 5.17. Proportional counter as used in WD spectrometers: an X-ray photon entering the window causes ionisation of the gas; the electric field around the anode wire causes multiplication of ions and electrons, giving an amplified output pulse proportional in height to the X-ray energy.

type, the source–crystal distance is fixed and interchangeable crystals are provided. Each of these has a radius of curvature that gives satisfactory results over a limited range of wavelengths. There is some cost saving compared with the fully focussing type, at the expense of sacrificing full wavelength coverage.

In the 'parallel-beam' type of WD spectrometer, the X-rays are collimated by means of focussing optics making use of the phenomenon of low-angle total reflection. The resulting parallel beam is reflected by a flat crystal (or multi-layer), giving higher intensities than in a conventional WD spectrometer for long wavelengths (the efficiency of the focussing optics decreases with decreasing wavelength). It is necessary for the specimen surface to lie in the correct plane relative to the focussing optics: in an SEM lacking an optical microscope, the appropriate location can be obtained by setting the spectrometer on a known X-ray line and adjusting the specimen height for maximum intensity.

5.3.4 Proportional counters

In a WD spectrometer X-rays are detected with a 'proportional counter' consisting of a gas-filled tube with a coaxial wire held at a positive potential between 1 and 2 kV (Fig. 5.17). Ionisation of the gas atoms by X-rays generates free electrons and positive ions, which are attracted respectively to the anode wire and to the body of the counter (acting as cathode). The accelerated electrons cause further ionisation, creating an 'avalanche', which results in a pulse of electrical charge appearing on the anode. The size of the pulse is dependent on the initial number of ions produced by the X-ray photon and, since this number is proportional to the energy of the absorbed photon, the pulse height is proportional to this energy. Electron multiplication in the counter gas is strongly dependent on the anode voltage: as this is varied the *absolute* pulse heights for all energies therefore change, while the *relative* heights retain a constant relationship.

Usually proportional counters are filled with argon (or sometimes xenon), with added methane (typically 10%). The X-rays enter through a 'window' which, for long wavelengths, must be so thin (to minimise absorption) that it is not completely gas-tight, necessitating a continuous supply of gas to compensate for leakage ('flow counter'). For shorter wavelengths a thicker, impervious, window can be employed, the counter being sealed for life ('sealed counter'). This type of counter is usually filled with xenon because of its greater absorption of short-wavelength X-rays, though sometimes an argon flow counter with increased gas pressure is used instead.

A 'tandem' configuration with a flow counter in front and a sealed counter behind (the former with an exit window as well as an entrance window) provides efficient detection of all wavelengths. This arrangement is especially appropriate for SEMs fitted with only one WD spectrometer. In multi-spectrometer instruments each spectrometer usually has a counter dedicated to either short or long wavelengths.

Pulse-height analysis

As described in the preceding section, the mean amplitude of the output pulses from a proportional counter is proportional to the energy of the detected X-ray photons. However, the ionisation of the gas is subject to statistical fluctuations, causing variations in height. Ideally the pulse-height distribution conforms to a Gaussian distribution function (Fig. 5.2), the width of which varies as $E^{0.5}$. However, broadening and asymmetry may occur because of contamination of the anode wire, which can be tolerated up to a point, but eventually it becomes necessary to replace the counter. Also, at high count-rates depression of the mean pulse height and broadening of the distribution may also occur owing to the high density of positive ions around the anode.

By applying 'pulse-height analysis', the first-order reflection of the crystal can be selected in preference to orders greater than 1, which have higher energy (by a factor n). This is achieved by use of a pulse-height analyser (PHA), which allows only pulses with heights lying within a certain 'window' to pass (Fig. 5.18). Sometimes, when there is no possibility of high-order reflections, a simple 'discriminator' is used: in this case all pulses lying above the threshold voltage are accepted. The need to match the PHA window to the pulses of interest is thus avoided, while unwanted low-amplitude noise is still suppressed.

The height of the output pulses from a proportional counter (for a particular X-ray energy) depends on the gas density. Pulse-height analysis is therefore sensitive to changes in temperature or pressure when using a flow counter (sealed counters are immune to such effects). The problem can be solved by using a gas-density stabiliser.

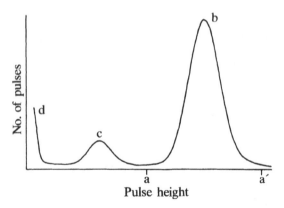

Fig. 5.18. The pulse-height distribution from a proportional counter: the main Gaussian peak (b) is accompanied by an escape peak (c) and electronic noise (d), which can be excluded by pulse-height analysis, whereby only pulses lying within 'window' aa' are accepted.

Escape peaks

It has been assumed so far that the whole energy of an X-ray photon absorbed in the counter gas is devoted to generating electrons and positive ions. There is, however, a finite possibility that a fluorescent Ar K photon (in the case of an argon-filled counter) emitted following absorption of an incoming X-ray photon may escape from the counter without being absorbed. This gives rise to an 'escape peak' in the pulse-height distribution, similar to that observed in ED spectra (Section 5.2.6) except that it is located the equivalent of 2.96 keV below the main peak (Fig. 5.18). This occurs only for X-rays of greater energy than the relevant excitation energy (3.20 keV for Ar). When the escape peak of a high-order reflection lies within the PHA window as set to accept a first-order reflection, pulse-height analysis is less effective (for example, see Pyle *et al.*, 2005).

5.3.5 Pulse counting and dead-time

In WD analysis, X-ray intensities are measured by counting the output pulses from the proportional counter, after amplification and selection by the PHA. For quantitative analysis the number of pulses arriving in a given time interval is counted, with the spectrometer set on each relevant peak. The selection of counting time is related to the count-rate and the total number of counts needed for the required statistical precision (see Section 8.4). For some purposes a continuous analogue intensity read-out provided by a 'ratemeter' is useful. In present-day instruments, this is presented on the computer screen, typically in the form of a 'thermometer' display.

The 'dead-time', defined as the time interval (t) after the arrival of a pulse during which the system is unresponsive to further pulses (typically a few microseconds) has the effect that the measured count-rate (n') is less than the true count-rate (n) by an amount that becomes significant at high count-rates. This is described by the equation

$$n' = n(1 - nt),$$ (5.4)

which differs from that applicable to ED systems (Section 5.2.4) because in this case the dead-time is 'non-extendable'. With a typical dead-time of $3\,\mu s$ the correction is 10% at $30\,000$ counts s^{-1}. Measured intensities can be corrected for dead-time by applying Eq. (5.4).

5.4 Comparison between ED and WD spectrometers

Count-rates per unit beam current for pure elements, as obtained with the crystals normally used in WD spectrometers, are plotted in Fig. 5.19. For a given crystal, the efficiency decreases with increasing wavelength, since the solid angle decreases with increasing Bragg angle. Count-rates obtained with WD spectrometers are considerably lower than those given by a typical ED

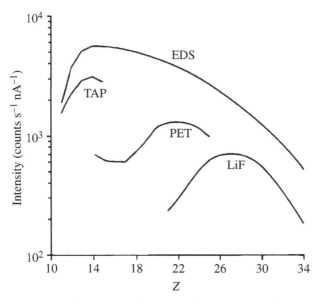

Fig. 5.19. Intensities of Kα lines for pure elements (atomic number Z) as recorded by an ED spectrometer and a WD spectrometer with different crystals (accelerating voltage 20 kV).

spectrometer, but this is counterbalanced by the better resolution of the former.

The peak-to-background ratio is related to resolution since the background count-rate depends on the width of the band of continuum which is recorded by the spectrometer. Typical values obtained with WD spectrometers range from a few hundred to over 1000 (for pure elements) and are approximately a factor of ten higher than for ED spectrometers, resulting in lower elemental detection limits (Section 8.5). WD spectrometers have the further advantage that there is more scope for increasing count-rates by increasing the beam current than there is with ED spectrometers, owing to the latter's limited maximum count-rate (for the whole spectrum).

Plate 1. Stereoscopic image (400 × 400 µm) of clinoptilite crystals (view with red/green spectacles).

Plates 1-8 are available for download in colour from www.cambridge.org/9780521142304

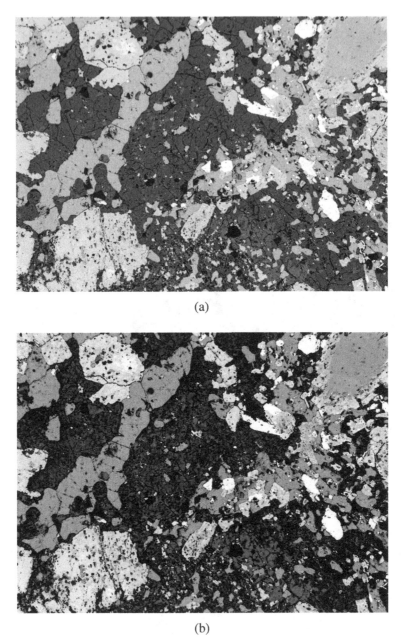

(a)

(b)

Plate 2. BSE images of igneous rock: (a) original monochrome image; (b) same image with grey scale converted to 'thermal' colours. (See Section 4.7.2.)

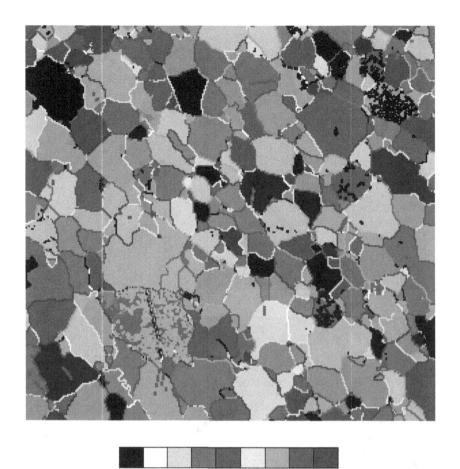

Plate 3. Orientation map of calcite grains in marble derived from EBSD patterns; misorientations across boundaries coloured according to scale shown (0–90°). (See Section 4.8.3.) (By courtesy of G. Lloyd.)

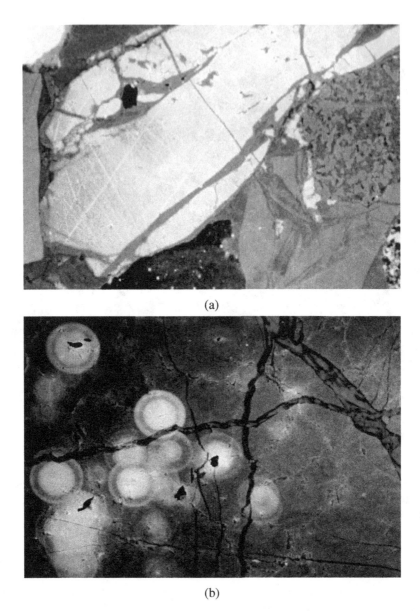

(a)

(b)

Plate 4. 'Real' colour SEM cathodoluminescence images obtained by combining images recorded with red, green and blue filters (See Section 4.8.4.): (a) quartz-cemented sandstone ($700 \times 500\,\mu m$) with two episodes of fracturing revealed by dark blue and red luminescent quartz infilling (Markowitz and Milliken, 2003); (b) haloes in quartz ($550 \times 350\,\mu m$) revealing radiation damage caused by radioactive elements in small inclusions (Oliveira *et al.*, 2003).

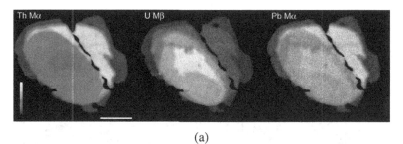

(a)

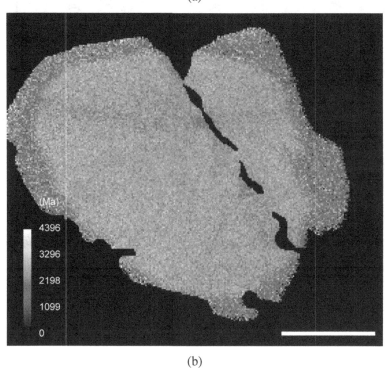

(b)

Plate 5. Monazite grain: (a) X-rays maps for Th, U and Pb; (b) map showing colour-coded age (Ma) derived from concentrations of the same elements (scale bar = 20 µm). (See Section 6.5.) (Goncalves, Williams and Jercinovic, 2005.)

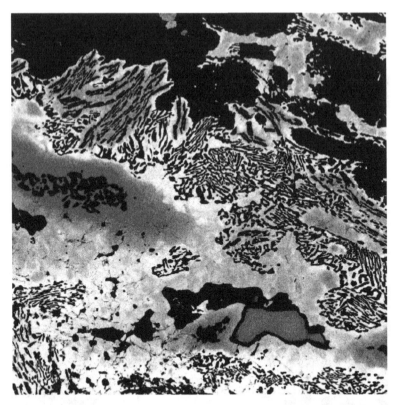

Plate 6. X-ray map of Ca in granulite, using 'thermal' colour scale, showing sillimanite (dark), zoned plagioclase, and symplectites (1×1 mm). (See Section 6.7.) (By courtesy of M. Jercinovic and M. Williams.)

Plate 7. Composite X-ray map (16×25 mm) of lunar meteorite consisting of regolith breccia matrix (bottom) and olivine-gabbro fragment (top); colours determined by amounts of Mg (red), Fe (green), and Ca (blue); principal phases: olivine – yellow, pigeonite – orange, augite – purple, feldspar – blue, various Fe-rich phases – green (Fagan *et al.*, 2003). (See Section 6.7.) (By courtesy of M. Killgore and T. Fagan.)

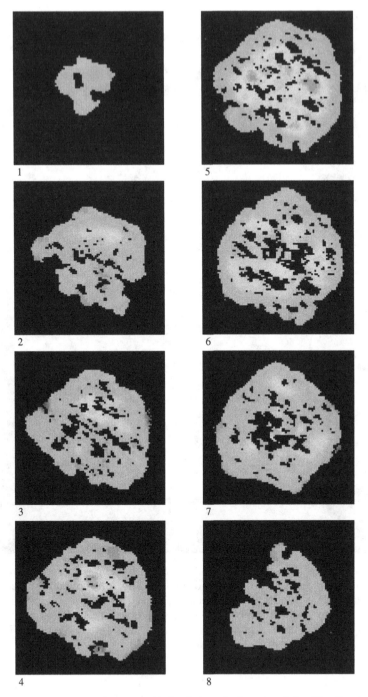

Plate 8. X-ray maps of garnet (450×450 μm) showing Mn distributions in serial sections (~40 μm slices); colour scale: blue-green-yellow-orange-red (Spear & Daniel, 1998). (See Section 6.10.) (By courtesy of F. S. Spear.)

6

Element mapping

6.1 Introduction

The spatial distribution of a specific element can be revealed by recording a 'map' of the intensity of its characteristic X-rays while the beam is scanned in a rectangular raster. A similar result can be obtained by leaving the beam position fixed while moving the specimen. Either ED or WD spectrometers can be used for X-ray mapping.

In the pre-digital era the normal form of X-ray image was the 'dot map', in which each recorded photon produced a bright dot on the display at a point corresponding to the position of the beam on the sample. The density of dots then showed variations in the concentration of the selected element. However, this approach has been superseded by digital mapping, which is described in the following section.

6.2 Digital mapping

With computer-controlled instruments X-ray maps are produced by recording the number of X-ray photons for a fixed time at each point in the scanned area and storing the data in the computer memory. A visible image is generated by converting this number into brightness on the screen (see Fig. 6.1). In its raw form the data consist of the number of X-ray counts recorded for each pixel. This can be converted into a standard type of image format in which the intensities are converted to 'grey levels' (typically 256). However, in this process information regarding absolute X-ray intensities, and thus concentrations, is lost. It is therefore desirable to save raw data for archival purposes in whatever form is provided, which generally will not be compatible with other software.

The relationship of dwell time, number of pixels and total acquisition time is shown in Fig. 6.2. While it is desirable to use a large number of pixels in the interests of image quality, this is pointless if it results in using such a short dwell

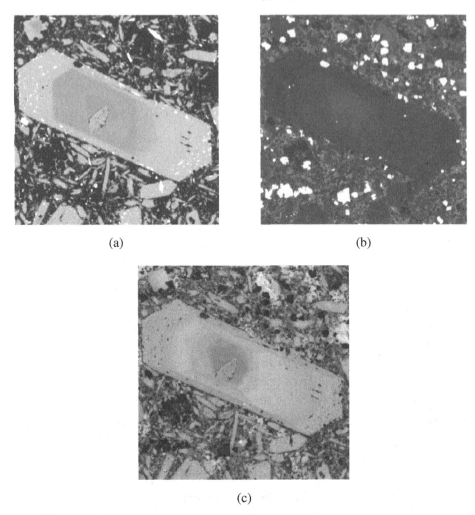

(a)   (b)

(c)

Fig. 6.1. X-ray maps of a clinopyroxene phenocryst: (a) Ca, (b) Fe and (c) Mg.

time per pixel that statistical fluctuations in the numbers of counts are large. Also, there is no advantage in using pixels smaller than the size of the X-ray source region (typically about 1 μm). In practice X-ray maps are usually limited to about 500 × 500 pixels, and the acquisition time required in order to obtain reasonable image quality is much greater than for electron images (reflecting the much lower yield of X-rays compared with backscattered or secondary electrons). Typical X-ray maps are shown in Fig. 6.1.

6.3 EDS mapping

An ED spectrometer collects the whole X-ray spectrum at once, but 'regions of interest' or energy bands containing the peaks of elements of interest can be

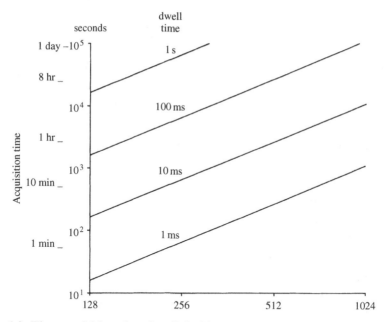

Fig. 6.2. The acquisition time for digital images versus number of pixels and dwell time per pixel.

defined and the outputs from these used for mapping. Owing to the effect of dead-time, ED systems are subject to count-rate limitations (Section 5.2.4) and the maximum rate obtainable is typically lower than for WD spectrometers, though the difference can be reduced by using a shorter pulse-processing time than usual (provided that the consequent sacrifice in energy resolution is not important). With a fixed dwell time per pixel, there is no allowance for dead-time, hence misleading results may be obtained at high count-rates, when the recorded count-rate decreases with increasing input count-rate (Fig. 5.5).

Owing to the broadness of the peaks in the ED spectrum, background is more important than for WD spectrometers, significant X-ray intensity being recorded even in regions containing none of the element concerned. Further, the background (continuum) intensity varies with mean atomic number, which may give a false impression of differences in concentration. Methods of correcting for this are discussed in Section 6.6.

The large memory available in contemporary computers makes it possible to collect and store complete spectra at each point. Element maps can then be produced from peak intensities extracted retrospectively, thereby avoiding the need to make prior decisions as to which elements to record.

6.4 WDS mapping

There are advantages in using WD spectrometers for mapping, including their better ability to resolve closely spaced spectral lines and the higher peak-to-background ratios obtained. In addition, their smaller dead-time allows higher count-rates, enabling less noisy images to be obtained in a given time (assuming that specimen-damage considerations allow the use of a high beam current).

A disadvantage of WD compared with ED spectrometers is the 'defocussing' effect that occurs when the beam is deflected from its normal position (Section 5.3.2) resulting in a fall-off in intensity at the edges of the image (unless the scanned area is small). One way to overcome this is to change the spectrometer setting as a function of the beam deflection. The line scan should preferably be oriented perpendicularly to the plane of the spectrometer so that the Bragg angle requires changing only between lines. Where there are several spectrometers arranged around the column, the orientation obviously cannot be simultaneously correct for all of them, but this difficulty can be resolved by applying an offset calculated separately for each spectrometer from the x and y displacements.

An alternative approach is to leave the beam in a fixed position and move the specimen stage instead. The fact that stage movement is relatively slow compared with beam deflection is unimportant given the typically long acquisition times used for X-ray mapping in order to minimise 'noise'. Computer-controlled stage movements as used in electron microprobes, but not all SEMs, are required for this mode of operation.

6.5 Quantitative mapping

If a map showing true concentrations is required, rather than one merely showing raw peak intensities, a correction for background is necessary. When an ED spectrometer is used, a 'continuum map' derived from a peak-free region of the spectrum can be recorded at the same time as the elemental maps. After applying a scaling factor to allow for the shape of the continuum, background can be subtracted from the peak intensities.

When WD spectrometers are used, background is less significant owing to the higher peak-to-background ratio, but if necessary a 'background map' can be recorded with the spectrometer offset from the peak, though this is time-consuming since it must be done separately from the recording of the peak-intensity map. Another possibility is to measure backgrounds with each spectrometer on standards that do not contain the elements of interest and then subtract these from the intensities recorded from the sample, with a correction for the dependence of the continuum intensity on mean atomic number.

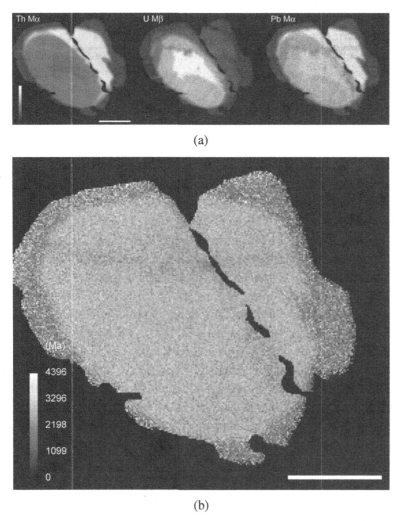

Fig. 6.3. (a) X-ray maps for Th, U and Pb in a monazite grain; and (b) a map showing the age (Ma) derived from concentrations of the same elements (scale bar = 20 µm). (Goncalves, Williams and Jercinovic, 2005.) See Plate 5 for colour version.

In ED spectra, overlap between neighbouring elemental lines occurs quite commonly (e.g. between Kβ and Kα lines of elements such as V and Cr, and between the L lines of heavy elements). This requires the spectrum recorded at each pixel to be processed in order to 'deconvolve' line overlaps, using methods described in Section 7.5.2.

For fully quantitative results it is necessary to apply matrix (ZAF) corrections (Section 7.7). To save computing time, the relatively simple 'alpha-coefficient' method (Section 7.7.4) can be used. The final outcome is a 'quantitative map' representing variations in true elemental concentrations.

In the case of phases containing U and Th, with their radioactive decay product Pb, concentrations of these elements can be converted into age values, which can be displayed in map form (Fig. 6.3). (Note that, in the absence of isotopic data, the ages are valid only if the concentration of non-radiogenic Pb is negligible.)

6.6 Statistics and noise in maps

X-ray photons are emitted randomly and the numbers recorded in a given time interval are subject to statistical fluctuations, which tend to obscure real intensity variations. For a 256×256-pixel X-ray map only 0.1 s counting time per pixel is available for an acquisition time of 2 h. With typical X-ray count-rates, statistical fluctuations therefore result in images that are significantly 'noisy'.

The scatter in the number n in an area of uniform brightness is proportional to $n^{0.5}$, and the relative fluctuation is thus $n^{0.5}/n$, or $n^{-0.5}$. Contrast (c) may be defined as $(n'-n)/n$, where n' is the number of X-ray counts in the object. The minimum contrast (c_{min}) consistent with visibility is proportional to the size of the statistical fluctuations and is thus given by $c_{min} = kn^{-0.5}$. From experiments on computer-generated images, Bright (1992) inferred that a value of 70 for k gives the minimum contrast required for detection of an object with near certainty. It follows that the larger its size, the more easily a low-contrast object can be detected: increasing the magnification thus makes such objects more visible.

6.7 Colour maps

The human eye can distinguish only about 16 different 'grey levels' in a monochrome image but far more different colours can be perceived, so it is advantageous to replace grey levels by 'false' colours. If too many colours are used the result tends to be confusing: a reasonably simple colour scale is therefore desirable. One that is commonly used is the 'thermal' scale, in which the colours correspond approximately to those of 'black-body' radiation, with concentration equated to temperature (e.g. black, purple, red, orange, yellow, white). An alternative is the 'rainbow' scale (violet, blue, green, yellow, red). An example of a false-colour X-ray image is shown in Plate 6.

Colours can be used to combine in one image information on several elements, each being assigned a different colour (Plate 7). Also, images can be created in which chemically distinct phases are represented by different uniform colours, determined by whether the intensities of selected X-ray lines are above or below certain threshold levels.

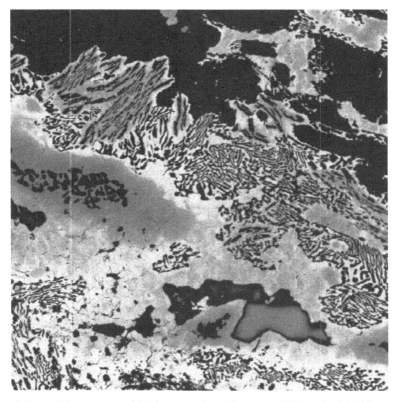

Fig. 6.4. An X-ray map of Ca in granulite, showing sillimanite (dark), zoned plagioclase, and symplectites (1 mm × 1 mm). (By courtesy of M. Jercinovic and M. Williams.) See Plate 6 for colour version.

6.8 Modal analysis

Modal analysis involves determining the volume fractions of constituent minerals in a rock from the relative areas measured on a planar surface, which traditionally has been done by point-counting using a microscope, the minerals being identified visually. Not only is this approach laborious but also some minerals are difficult to identify rapidly and reliably in the microscope. Furthermore, fine textures present problems, and opaque phases usually are not identified. Automated electron microprobe modal analysis overcomes most of these limitations.

In the case of grain mounts, the BSE signal can be used to indicate the presence of a grain, before making X-ray measurements. ED point analyses can be carried out over the whole area of each grain, so that not only mineral volume fractions but also mean compositions, variances, etc. can be determined.

Similar procedures can also be used for 'microprospecting' for rare phases such as gold, whereby large areas are covered rapidly, and X-ray analysis is

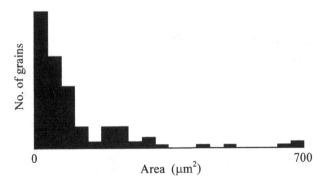

Fig. 6.5. A histogram showing the size distribution of grains in Fig. 4.30(b), obtained using image-analysis software.

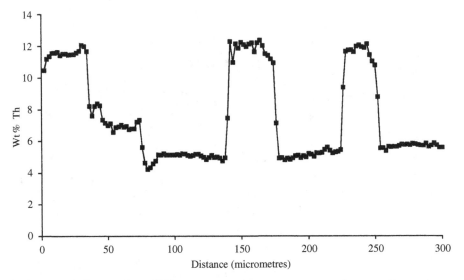

Fig. 6.6. A linear plot of Th concentration across a zoned monazite grain, obtained by moving the specimen stage in 2-μm steps.

carried out where the BSE signal indicates the possible presence of a grain of interest. A related application is the determination of the ore content in exploration drill cores or tailings from extraction processes (Reid *et al.*, 1985). A low-vacuum SEM system developed specially for use in mineral exploration has been described by Robinson (1998).

Similar approaches can be applied to BSE images (where there are enough atomic-number differences to discriminate between phases), for which purpose standard image-analysis software can be used. Information about grain-size distribution (see Fig. 6.5), volume fraction, shape, and mineral associations can be obtained. This can be useful in fields such as mineral processing (for example, see Lastra, Petruk and Wilson, 1998).

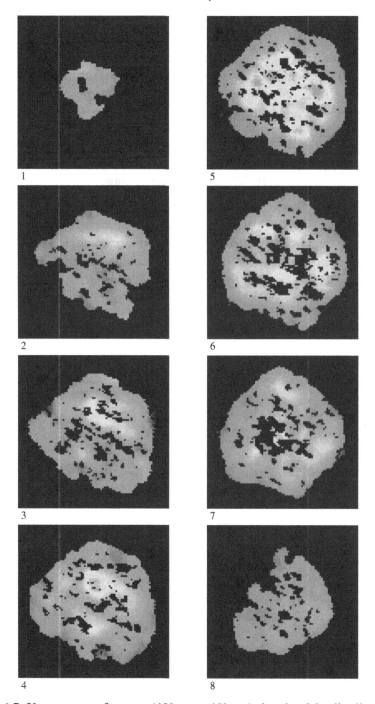

Fig. 6.7. X-ray maps of garnet (450 μm × 450 μm) showing Mn distributions in serial sections (∼40-μm slices); (Spear and Daniel, 1998). (By courtesy of F. S. Spear.) See Plate 8 for colour version.

6.9 Line scans

As noted in Section 6.6, X-ray maps are prone to statistical 'noise', owing to the short dwell time per pixel which is available if the total acquisition time is to be kept within reasonable limits. However, if the X-ray intensity is plotted while the beam is scanned along a single line, a less noisy result can be obtained in a much shorter time (see Fig. 6.6 for an example).

6.10 Three-dimensional maps

Element distributions in conventional maps are representative of the composition at the surface of the specimen (or, strictly speaking, within a depth of approximately 1 μm from the surface). Element distributions in three dimensions can, however, be reconstructed from a series of maps, using a serial sectioning technique in which a controlled thickness is removed by grinding and the surface is repolished at each stage (Fig. 6.7, Plate 8; Spear and Daniel, 1998).

7

X-ray analysis (1)

7.1 Introduction

Qualitative X-ray analysis entails the identification of the elements present in a given sample, or the identification of phases from the elements which they contain. Of the two available sorts of X-ray spectrometer, the ED type is far better for qualitative analysis, owing to its ability to record complete spectra rapidly (major elements and their approximate relative concentrations are apparent in only a few seconds). Occasionally, though, ambiguity in the identification of peaks closely similar in energy requires the better resolution obtainable with a WD spectrometer.

For quantitative analysis X-ray line intensities emitted from the specimen are measured and elemental concentrations are calculated from the ratios of these intensities to those from standard samples with known concentrations. Methods of measuring intensities and correcting for background differ in WD and ED analysis, which are therefore treated separately below. The 'matrix' (or 'ZAF') corrections required in order to allow for the effect of the difference in composition between standard and specimen on the emitted intensities are common to both methods of analysis.

7.2 Pure-element X-ray spectra

The origin of characteristic X-rays is described in Section 2.5.2. The K lines of elements of atomic number 11–30 (Na–Zn) lie within the energy range 1–10 keV. Elements of higher atomic number can be identified from their L or M lines, which lie in the same range. With a thin-window ED detector (Section 5.2.1), the atomic-number range can be extended down to 4 (Be). The full range of atomic numbers from 4 to 92 is available using WD spectrometers, with suitable choice of crystal and counter (Section 5.3).

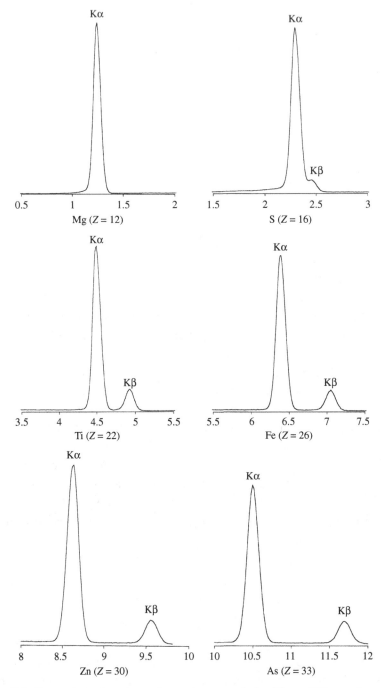

Fig. 7.1. K spectra of various elements, recorded with an ED spectrometer (energy in keV), showing the dependence of both the energy of the Kα peak and the relative intensity and position of the Kβ peak on atomic number (Z).

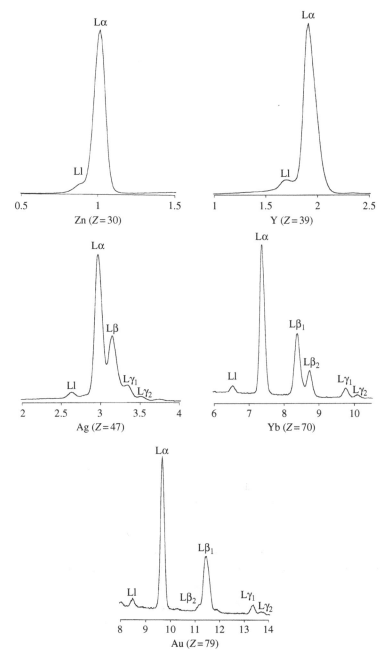

Fig. 7.2. L spectra of various elements, recorded with an ED spectrometer (energy in keV), showing the dependence of both the energy and the complexity of the spectra on atomic number (Z).

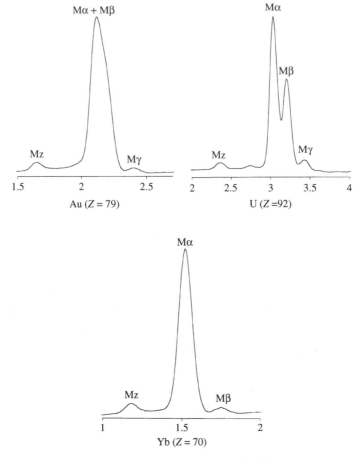

Fig. 7.3. M spectra of various elements, recorded with an ED spectrometer (energy in keV), showing the dependence of both the energy and separation of Mα and Mβ peaks on atomic number (Z).

K spectra as recorded by EDS contain a maximum of two lines per element (Kα and Kβ). For low atomic numbers the Kβ line is either completely merged with the Kα peak or appears as a 'shoulder' on the high-energy side (see Fig. 7.1). Similar behaviour occurs with L spectra, except that six or more lines may be visible (Fig. 7.2). M spectra of heavy elements contain fewer discrete lines (Fig. 7.3).

7.3 Element identification

X-ray lines may be identified by reference to tables of energies or wavelengths. However, various aids are available, including on-screen line markers and

(a)

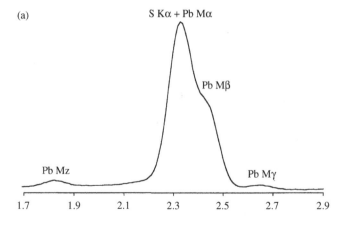

(b)

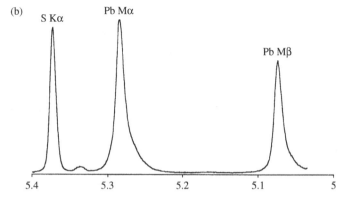

Fig. 7.4. (a) The ED spectrum of galena, showing unresolved Pb and S peaks, and (b) the WD spectrum, showing well-resolved Pb and S lines (wavelength in ångström units; scale reversed for compatibility with energy scale).

automatic identification (the results of which should, however, be checked for plausibility). Situations in which there is ambiguity between α lines of different elements are rare. Cases of lines being unresolved in the ED spectrum are commoner (for example S Kα / Pb Mα, Ti Kα / Ba Lα, Si Kα / Sr Lα and P Kα / Zr Lα), but usually can be recognised from the non-Gaussian shape of the combined peak (see Fig. 7.4(a)).

WD spectra are similar to ED spectra except that the lines are sharper and traditionally are plotted versus wavelength, so that the lines occur in reverse order. More minor peaks are visible, owing to the higher peak-to-background ratio and better resolution. The same principles of peak identification as described above can be applied. Figure 7.4(b) illustrates the use of WDS to resolve lines that overlap in the ED spectrum.

The presence of high-order lines (Section 5.3.1) is a complicating factor in WD spectra. They can be suppressed by means of pulse-height analysis, though not always completely (see Section 5.3.4).

7.4 Mineral identification

The main object of finding out which elements are present is usually to identify the phase. Common rock-forming minerals contain principally the following major elements: Na, Mg, Al, Si, S, P, K, Ca, Ti, Cr, Mn and Fe (also O, which is detectable only with a thin-window ED detector, however). For sulphides and related phases additional elements are relevant, e.g. Co, Ni, Cu, Zn, Pb and sometimes other elements such as As, Se, Sb and Bi. For carbonates C, like O, is detectable only with a thin-window detector.

Examples of ED spectra of common minerals are given in the appendix. Different minerals containing the same major elements in different proportions are often distinguishable on the basis of the relative heights of the peaks (e.g. for pyroxenes the Si peak is larger relative to Fe and Mg than for olivines).

Mineral identification based on electron microprobe data is facilitated by using a database such as that described by Smith and Leibowitz (1986). Automatic mineral identification is useful in modal analysis (Section 6.8). For limited groups of minerals in specific types of rocks quite simple methods of mineral identification are adequate. For example, Nicholls and Stout (1986) ranked major elements in silicates in order of peak intensity as the basis for identification, supplemented by intensity ratios when necessary.

7.5 Quantitative WD analysis

The electron microprobe is designed for quantitative analysis as its principal function, for which WD spectrometers are normally used, as described in the present section. An SEM equipped with a WD spectrometer is less satisfactory for this purpose in some respects. More commonly SEMs have only an ED spectrometer, which can be used quantitatively, with some restrictions, as discussed in the next section.

When using WD spectrometers, a suitable crystal must be chosen for each line (Section 5.3.1) and appropriate conditions for pulse-height analysis selected (Section 5.3.4). The measured intensities need to be corrected for dead-time (Section 5.3.5). The beam current should be chosen to be of such a value as to avoid count-rates above about 5×10^4 counts per second, which could cause pulse-height depression and unduly large dead-time corrections. For trace-element analysis using a high current, excessive count-rates on the standards can be avoided by using a lower current and adjusting the intensities proportionally. Counting times are chosen according to criteria discussed in Section 8.3.

The spectrometer should be set accurately to the peak position by executing a 'peak-seek' procedure on the standard. As a rule this need not be repeated in the course of a series of measurements unless these extend over many hours. The surface of the specimen should always be in the correct plane, to ensure that a constant Bragg angle is maintained: with vertically mounted spectrometers this can be achieved by adjustment of the specimen height to obtain a sharp image in the optical microscope. Scanning electron microscopes are not equipped with such a microscope, but WD spectrometers mounted in the inclined orientation are relatively immune to the effect of specimen-height differences (see Section 5.3.2).

It is important that the beam current does not change during the course of intensity measurements on specimen and standards. Current drift should not be significant in an instrument with beam-current regulation, but as an extra safeguard the current can be monitored before each measurement, and the X-ray intensities normalised. In the absence of a Faraday cup, drift can be taken into account by repeated measurement of the X-ray intensity from a standard.

X-ray intensities are strongly dependent on the incident electron energy, E_0, which is governed by the electron accelerating voltage. The effective E_0 value may be reduced as a result of surface charging in the case of a non-conducting specimen with an inadequate conductive coating, causing a reduction in X-ray intensity and consequently bad analytical results. This effect is observable in the continuous X-ray spectrum as a lowering of the upper cut-off or Duane–Hunt limit (see Section 2.5.1). Charging can also be diagnosed from instability in secondary-electron images (Section 4.6.2), or low and unstable absorbed current readings. For accurate quantitative analysis such effects must be eliminated by recoating or using conducting paint to ensure good contact between specimen and holder.

It can usually be assumed that peak positions for specimen and standard are the same. Slight shifts related to chemical bonding may occur, but can be avoided by suitable choice of a composite standard (e.g. Al_2O_3 for Al in silicates, rather than the pure metal). Chemical effects are more marked for elements of atomic number below 10, which require special procedures (see Section 8.1). Small peak shifts, whether caused by chemical effects or other factors (e.g. variations in specimen height), can be taken into account by combining intensities recorded at two positions, one on each side of the peak, or at several points across the peak.

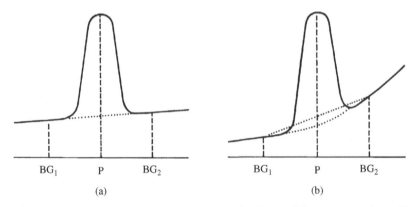

Fig. 7.5. Background determined by averaging intensities measured at offset positions (BG_1 and BG_2): (a) linear background – the correct result is obtained for the background under the peak; (b) curved background – an incorrect result is obtained.

7.5.1 Background corrections

The measured peak intensity includes a contribution from background originating mainly from the X-ray continuum (Section 2.5.1). The background varies only slowly as a function of wavelength and can be estimated by taking the average of measurements on each side of the peak (Fig. 7.5). Background curvature is negligible for most purposes but can be significant in the region of the tail of a neighbouring major peak (Fig. 7.5(b)). If background measurements are increased by overlap from a neighbouring peak, the apparent peak-minus-background intensity is too low and may even be negative. This effect can be avoided by measuring background on one side only, where there is no overlap (but slope must then be taken into consideration).

In choosing background positions one should avoid the absorption edges of major elements present in the sample, which cause a step in the continuum. Also, when an argon-filled counter is used the step at $3.870\,\text{Å}$ due to the Ar K absorption edge should be taken into consideration.

7.5.2 Overlap corrections

Sometimes peak intensities are enhanced by overlap from a neighbouring peak, requiring an 'overlap correction' (Fig. 7.6). In the approach described by Fialin (1992), the tail of the adjacent peak is represented by a second-order polynomial. Alternatively, the amount of overlap can be expressed as a fraction of the intensity of the principal peak of the overlapping element. Thus, in a specimen containing Ti and V, the contribution of Ti $K\beta$ to the measured

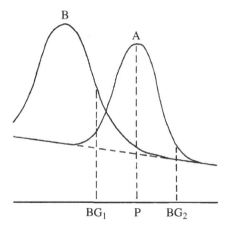

Fig. 7.6. Peak overlap: measurements made on peak A are influenced by the tail of peak B; BG_1 is most affected, causing the apparent peak-minus-background intensity to be negative.

V $K\alpha$ intensity is assumed to be a constant fraction of the Ti $K\alpha$ intensity. This approach can be used, for example, to correct for overlaps between rare-earth elements (Roeder, 1985). 'Overlap factors' can be determined empirically from pure-element standards, preferably with a correction for matrix effects (Donovan, Snyder and Rivers, 1993). For accurate trace-element analysis, careful investigation of all possible overlaps is necessary (see, for example, Jercinovic and Williams, 2005).

An approximate correction can be applied retrospectively: if a spurious concentration of $x\%$ of element X is found in pure Y, overlap in an 'unknown' sample can be corrected by multiplying the concentration of Y by $x/100$ and subtracting the result from the apparent concentration of X. Such corrections may be either negative or positive, depending on whether peak or background measurements are affected most.

7.5.3 *Uncorrected concentrations*

According to the 'Castaing approximation', the intensity of a characteristic X-ray line is proportional to the mass concentration of the element concerned. A first estimate for the concentration of a given element (A) in an 'unknown' specimen is thus obtained from the following expression:

$$C'_A(\text{sp}) = [I_A(\text{sp})/I_A(\text{st})]C_A(\text{st}), \qquad (7.1)$$

where $I_A(\text{sp})$ and $I_A(\text{st})$ are the intensities measured on specimen and standard respectively, $C_A(\text{st})$ is the mass concentration of 'A' in the standard, and $C'_A(\text{sp})$ is the 'uncorrected concentration' in the specimen. The ratio $I_A(\text{sp})/$

Table 7.1. Olivine analysis (accelerating voltage 15 kV; Kα lines; oxygen estimated by difference)

Element	Standard	Specimen peak counts	Standard peak counts	Concentration (wt%) in standard	Uncorrected concentration (wt%) in specimen	Specimen matrix factor	Standard matrix factor	Corrected concentration (wt%) in specimen
Mg	MgO	121 743	270 889	60.3	27.1	0.709	0.791	30.2
Si	CaSiO$_3$	85 421	184 273	34.3	15.9	0.715	0.862	19.2
Fe	Fe	5 512	98 429	100.0	5.6	0.819	1.000	6.8

$I_A(st)$ is commonly known as the 'k ratio'. Standards may be pure elements or compounds (see Section 7.10). An example of the calculation of corrected concentrations is given in Table 7.1. The origins of the 'matrix effects' which cause divergence from the Castaing approximation are described in Section 7.7.

7.6 Quantitative ED analysis

Though WD spectrometers are more commonly used for quantitative analysis, there are some advantages in using an ED spectrometer, as discussed below, and most SEMs are equipped solely with this type. The essentials are the same as described above, but there are significant differences.

7.6.1 Background corrections in ED analysis

Background is more important in ED than in WD analysis, because peak-to-background ratios are substantially lower. It may be measured by summing the contents of a number of channels in a peak-free region. Instead of linear interpolation, which is less reliable than for WD spectra, owing to the greater energy range involved, a preferable approach is 'continuum modelling', whereby the Kramers expression (Eq. (2.3)), with additional factors to allow for absorption in the sample itself and in the detector window etc., is fitted to peak-free regions of the spectrum (Fig. 7.7).

Alternatively, background may be removed by mathematical filtering using a 'top-hat' filter function, with a value $+1$ in the central zone and -1 in the 'wings' (Fig. 7.8). The filter is stepped one channel at a time through the spectrum and the sum of the products of the filter function and the contents of the relevant channels is calculated. The output is zero where there are no peaks, hence background is eliminated. This approach is especially suitable for cases such as particles and rough specimens, for which modelling the continuum is difficult.

7.6.2 Measuring peak intensities in ED analysis

For measuring ED peak intensities it is desirable to make use of as much of the peak profile as possible, in order to maximise the statistical precision. The simplest way of doing this is to sum the contents of all the channels within a 'region of interest' (ROI), the optimum width of which is approximately equal to the FWHM. Alternatively, a Gaussian function may be fitted to each peak and its area determined by integration. Overlap is taken into account automatically by finding the overall best fit for the whole spectrum. Experimental

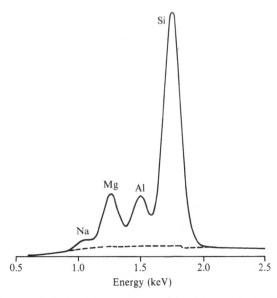

Fig. 7.7. Background derived by continuum modelling (dashed line): intensity beneath the peaks is obtained by interpolation from peak-free regions using a mathematical expression for the continuum.

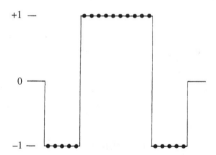

Fig. 7.8. 'Top-hat' digital filter used to distinguish peaks from background in ED spectra (see text).

spectra obtained from standards can be used instead for fitting purposes. For each element, the part of the spectrum containing not only the α line but also all others belonging to the same shell should be fitted. If 'top-hat' filtering is used to eliminate background, spectrum fitting may be carried out using filtered spectra both for the specimen and for standards.

7.6.3 Comparison between ED and WD analysis

The accuracy of quantitative ED analysis is comparable to that of WD analysis, at least for concentrations above 1%, provided that proper standards are used rather than a 'standardless' procedure (Section 7.10.1). However, detection limits are higher (typically around 0.1%). Also, peak overlaps are more common, owing to the relatively poor spectral resolution, though the methods of spectrum processing described above cope well with overlaps that are not too severe. For many applications (e.g. major elements in rock-forming silicates) quantitative ED analysis is perfectly satisfactory, and even has some advantages, including simpler setting up and compatibility with lower beam currents that minimise damage to carbonates, feldspars, glasses, etc.

The ideal instrument for quantitative analysis by either method is an electron microprobe, which has several advantages over the SEM, including beam-current regulation, the availability of an optical microscope and the use of an automated specimen stage. For WD analysis it is also a considerable advantage to have several spectrometers. Sometimes there is merit in combining the two methods – ED for major elements and WD for minor and trace elements (Ware, 1991). This requires the sensitivity of the ED spectrometer to be reduced so that a sufficiently high beam current for WD analysis can be used (see Section 5.2.3).

7.7 Matrix corrections

Matrix corrections are applied to uncorrected concentrations in order to obtain 'true' concentrations. The corrections may be expressed as separate factors represented by the acronym 'ZAF' (atomic number – absorption – fluorescence). Since these factors are dependent on the composition of the specimen, which is not known until the corrections have been calculated, an iterative procedure is used (Section 7.8). ZAF factors are also required for the standards and are calculated from their known compositions. Individual corrections are discussed in the following sections.

7.7.1 Atomic-number corrections

The efficiency with which characteristic X-rays are excited depends on the mean atomic number of the specimen, owing to two distinct phenomena – electron penetration and backscattering. The penetration of incident electrons is determined by the 'stopping power' of the specimen (Section 2.2), which decreases with increasing Z. The X-ray intensity generated (per unit concentration) is dependent on the mass penetrated and thus increases with Z. The correction for the loss of X-ray intensity due to electron backscattering is closely related to the electron backscattering factor η (the fraction of incident electrons that leave the specimen), which increases rapidly with Z (see Fig. 2.4). In the

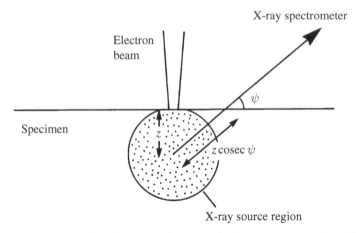

Fig. 7.9. X-rays produced at depth z travel distance $z \cosec \psi$ within the specimen before emerging (ψ is the X-ray take-off angle).

combined 'atomic-number correction', these two effects tend to counterbalance each other, but the backscattering term is dominant. Hence, when the sample has a higher mean atomic number than the standard the concentration must be corrected upwards (and vice versa).

7.7.2 Absorption corrections

X-rays travel a certain distance within the specimen before emerging. The resulting absorption depends on 'χ', defined as $\mu \cosec \psi$, where μ is the mass absorption coefficient of the specimen for the X-rays concerned (see Section 2.6) and ψ is the X-ray take-off angle (Fig. 7.9). For X-rays produced at depth z, the factor by which the intensity is reduced by absorption is $\exp(-\chi \rho z)$, where ρ is the density. In reality the X-rays are produced over a range of depths, with a distribution described by the function $\phi(\rho z)$. The overall factor by which the X-ray intensity is reduced is obtained by integrating $\phi(\rho z) \exp(-\chi \rho z)$.

The depth distribution function, $\phi(\rho z)$, is defined as the intensity generated in a thin layer at depth z, relative to that generated in an isolated layer of the same thickness, and takes the form shown in Fig. 7.10. In the classical ZAF correction method, $\phi(\rho z)$ is represented by a simple approximate expression, which gives satisfactory results provided that the absorption factor is not less than about 0.5. Alternative correction procedures based on relatively complicated but more realistic $\phi(\rho z)$ expressions, known collectively as 'phi–rho–z' methods (Heinrich and Newbury, 1991), are preferable. There are several correction algorithms, which give similar (but not identical) results. The

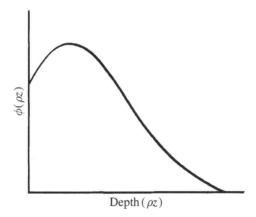

Fig. 7.10. The 'phi–rho–z' function representing the depth distribution of X-ray production.

'CITZAF' software package (Armstrong, 1995) offers a wide range of options, some of which may perform better for particular types of sample than others.

Absorption coefficients tend to increase as the wavelength of the absorbed X-rays increases. In silicates the corrections for Na, Mg and Al are thus quite large. Heavy elements also absorb strongly. The presence of absorption edges (Section 2.6) has a significant effect on the size of the correction in specific cases.

Geological samples are, of course, commonly non-conducting and the possibility that, even with a conducting surface coating, a build-up of negative charge in the region penetrated by the incident electrons could affect $\phi(\rho z)$ must be considered. There is little direct evidence for this and conventional correction procedures are used routinely for such samples with apparent success. However, Fialin (1988) attributed discrepancies for elements of atomic number 11–14 (for which absorption corrections are large) to charging, and proposed modifications to the correction procedure to take this into account.

7.7.3 Fluorescence corrections

The emission of characteristic X-rays of a given element can be excited by other X-rays when the energy of the latter exceeds the critical excitation energy of the former (Section 2.7). Fluorescence is excited by characteristic lines of other elements that satisfy the energy criterion (Fig. 7.11). For example, Fe Kα X-rays (6.40 keV) excite Cr (critical excitation energy 5.99 keV), but Cr Kα X-rays (5.41 keV) do not excite Fe (critical excitation energy 7.11 keV). A fluorescence correction is therefore required for the former, but not for the latter. For most geological samples characteristic fluorescence corrections are fairly small.

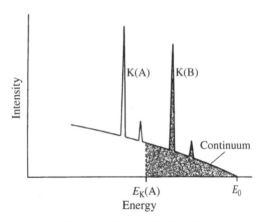

Fig. 7.11. Fluorescence excitation: X-rays in the shaded part of the spectrum have energy greater than the excitation energy, $E_K(A)$, of element A, therefore both the relevant part of the continuum and the characteristic radiation of element B can excite fluorescence.

Fluorescence caused by the part of the continuum above the critical excitation energy of the element of interest is always present, but the enhancement rarely exceeds 5%. Furthermore, it occurs in both specimen and standard, and so tends to cancel out. In principle a correction should be applied, but it is neglected in most correction programs.

Boundary fluorescence

The volume within which fluorescence is excited is considerably greater than that within which X-rays are generated directly by the electron beam, because the exciting X-rays are more penetrating. In deriving fluorescence corrections it is assumed that the composition of the former volume is the same as the latter – i.e. the specimen is homogeneous on the appropriate distance scale. If this is not so, the results of quantitative analysis may be in error.

The worst situation is when the fluoresced volume includes a high concentration of an element absent from the region penetrated by the bombarding electrons but strongly excited by X-rays emitted from this volume. In Fig. 7.12, Ti in ilmenite is excited by Fe in adjacent haematite, giving a spurious Ti concentration of about 1% when the beam is close to the boundary, decreasing exponentially with the distance of the beam from the boundary. In the case of olivine adjacent to Ca-rich pyroxene, spurious Ca concentrations of more than 0.1% may occur in the olivine (Dalton and Lane, 1996), necessitating a correction for olivine–clinopyroxene geothermobarometry (Llovet and Galan, 2003). Similarly, false Ti is observed in garnet adjacent to ilmenite (Feenstra and Engi, 1998).

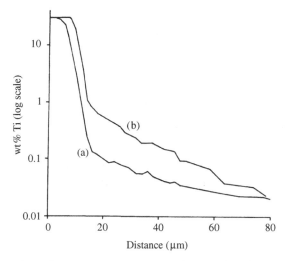

Fig. 7.12. Boundary fluorescence: apparent Ti concentration as a function of
the distance of the beam from a vertical boundary between ilmenite and
(a) glass, with continuum fluorescence excitation only; and (b) haematite,
with excitation by both continuum and Fe K radiation. (Accelerating voltage
20 kV; X-ray take-off angle 75°.) After Maaskant and Kaper (1991).

7.7.4 Alpha coefficients

The calculation of matrix correction factors can be greatly simplified by
assuming that the overall correction factor is given by $\Sigma c_i \alpha_i$, where α_i is the
'alpha coefficient' of element i, representing the influence which this element
has on the X-ray intensity of the element concerned. The summation is carried
out for all elements present. Sets of alpha coefficients have been compiled for
various conditions (Bence and Albee, 1968; Albee and Ray, 1970), but more
flexible methods based on physical models, as described previously, are now
generally preferred.

7.7.5 The accuracy of matrix corrections

If standards closely resembling the analysed sample are used, matrix corrections are small and uncertainties in them can be neglected. However, in
practice this condition is often not satisfied because of difficulties in finding
suitable standards, in which case matrix corrections may be quite large.
Uncertainties in the absorption correction (usually the largest) may be minimised by choosing a low accelerating voltage.

Matrix corrections are sensitive to the incident electron energy, E_0, as
determined by the electron accelerating voltage, and the value used in the

calculations should be accurate, preferably to within ± 0.1 keV. The simplest way to check the actual value is to observe the Duane–Hunt limit in the continuous X-ray spectrum (Section 2.5.1).

Provided that excessive absorption is avoided, errors due to matrix corrections are typically less than $\pm 2\%$. In some SEMs the specimen must be tilted for X-ray analysis: even with a suitably modified correction (Section 8.9.1), there is more uncertainty in the results than with normal incidence. Errors may be larger for extreme compositions, especially when these involve very heavy or light elements. Systematic errors have been noted, for example, in the determination of Ag in gold (Reid, le Roex and Minter, 1988) and S in cinnabar (Harris, 1990).

7.8 Correction programs

To calculate correction factors, a composition must be assumed, for which uncorrected concentrations can be used as an initial approximation. Having obtained a set of correction factors, updated concentrations are derived and used to recalculate the correction factors. When there are no further significant changes in the concentrations this iterative process is terminated.

7.8.1 Unanalysed elements

Quite commonly not all of the elements present in the sample are included in the analysis, the most obvious example being oxygen in silicates (the extra time required to determine oxygen directly renders it not worthwhile for most purposes). Since correction calculations require full knowledge of the composition, some assumption about such 'missing' elements must be made. The oxygen concentration may be obtained either from stoichiometry or 'by difference', that is, subtracting the sum of the concentrations of all other elements from 100%. In the former case an appropriate amount of oxygen is allocated to each cation according to its valency. However, difficulties arise with multivalent elements, notably Fe. The difference method is preferable in this respect and for dealing with water.

Available software does not always make provision for more than one unanalysed element. For carbonates, calculating O by difference (implying that C is equivalent to O) has only a minor effect on the results. Very similar results are obtained using the procedure of Lane and Dalton (1994), whereby four O atoms are assigned to each (divalent) metal atom by assuming a valency of eight, giving MO_4 in place of MCO_3. The effect of H in hydrous phases can safely be neglected.

7.9 Treatment of results

The result of a quantitative electron microprobe analysis is expressed in the first instance as elemental mass concentrations (weight per cent). The concentrations of unanalysed elements such as O are obtained by calculation, as just described. For silicates, etc., it is usual to express concentrations as weight per cent of the appropriate oxide (see Table 7.2). Iron is usually given as FeO, though in some minerals (e.g. aegirine, andradite, epidote, scapolite, serpentine and sodalite) it occurs as Fe_2O_3 and in others it is present in both forms (see below). In the case of sulphides, elemental concentrations only are required.

For many minerals the oxide sum should be close to 100% (between 99% and 101% is acceptable for most purposes). A low total can be caused by beam-current drift, poor spectrometer calibration, etc., but may occur for other reasons, such as the presence of water or an element not included in the analysis. Normalisation to 100% is undesirable because it disguises these effects.

Another cause of low totals is the assumption that Fe is divalent when some or all is actually in the trivalent state: for example, an analysis of magnetite (Fe_3O_4) gives 93.1% FeO if Fe^{3+} is neglected. Conversely, totals exceeding 100% may occur if Fe^{2+} is incorrectly assumed to be Fe^{3+}. A high total can also arise when an anionic element such as F or Cl substitutes for O. This may be corrected by deducting the appropriate amount of oxygen given by $16C_X/A_X$, where C_X is the mass concentration and A_X the relative atomic mass of the substituting element.

Table 7.2. *Normal valencies and oxide ratios (wt% oxide/wt% element) for common elements*

Element	Valency	Oxide	Oxide/ element ratio	Element	Valency	Oxide	Oxide/ element ratio
Na	1	Na_2O	1.348	Fe	2	FeO	1.286
Mg	2	MgO	1.658	Fe	3	Fe_2O_3	1.430
Al	3	Al_2O_3	1.890	Ni	2	NiO	1.273
Si	4	SiO_2	2.139	Rb	1	Rb_2O	1.094
P	5	P_2O_5	2.291	Sr	2	SrO	1.183
K	1	K_2O	1.205	Y	3	Y_2O_3	1.270
Ca	2	CaO	1.399	Zr	4	ZrO_2	1.351
Ti	4	TiO_2	1.668	Ba	2	BaO	1.117
V	3	V_2O_3	1.471	La	3	La_2O_3	1.173
Cr	3	Cr_2O_3	1.461	Pb	2	PbO	1.077
Mn	2	MnO	1.291	U	4	UO_2	1.134

Table 7.3. *pyroxene analysis*

Element	Concentration (wt%) of element	Concentration (wt%) of oxide	Atom per cent	Atoms per six O
Si	24.64	52.70	19.57	1.962
Ti	0.20	0.34	0.09	0.009
Al	0.97	1.84	0.81	0.081
Fe	5.82	7.33[a]	2.32	0.233
Mn	0.12	0.16	0.05	0.005
Mg	9.14	15.15	8.39	0.841
Ca	15.43	21.58	8.59	0.861
Na	0.36	0.49	0.35	0.035
O	43.32[b]			
Total	100.00	99.59	59.83	4.027

[a] Calculated as FeO.
[b] Calculated by difference.

Table 7.3 shows a typical silicate analysis. Concentrations are usually given to two decimal places, but this should not be taken to indicate *accuracy*. Errors given in results tables are usually those estimated from counting statistics only. More figures after the decimal point are needed for trace elements. 'Numbers of atoms' are normalised with respect to six oxygens (which is appropriate for pyroxene), oxygen being assigned to each element according to its valency. The cation total is close to the theoretical value of four for pyroxene, this being a useful internal test of the quality of the analysis. The cations can also be assigned to different lattice sites, to give a structural formula, as described below. Calculating to the ideal cation total is sometimes preferred.

7.9.1 Polyvalency

The Fe^{2+}/Fe^{3+} ratio can be calculated from EMPA data by allocating O atoms estimated by difference first to the other (monovalent) cations and dividing the remainder between Fe^{2+} and Fe^{3+}, but the accuracy of the results is less than is desirable for geothermobarometry (Cosca, Essene and Bowman, 1991; Schumacher, 1991). In principle more accurate results may be obtained by using measured O concentrations (e.g. Herd, Papike and Brearley, 2001), but the accuracy achievable for such light elements is limited (see Section 8.1). Calculation methods for Fe^{2+} and Fe^{3+} in silicates have been discussed by Droop (1987). For garnets and pyroxenes calculation is straightforward, but assumptions about site occupancy are necessary for amphiboles (Jacobson, 1989). Micas are problematic.

Table 7.4. *Numbers of oxygen atoms and total cations in formulae of common minerals*

Mineral	No. of O atoms	No. of cations
Amphibole	23	15
Chlorite	28	20
Cordierite	18	11
Epidote	25	16
Feldspar	8	5
Garnet	24	16
Ilmenite	3	2
Kaolinite	18	8
Kyanite	5	3
Mica	22	16
Mullite	13	8
Nepheline	16	12
Olivine	4	3
Pyroxene	6	4
Spinel	4	3

X-ray spectra are to a large extent free of chemical effects. However, the wavelength of the S Kα line shows relatively large differences with valency, which can be utilised to distinguish between sulphates and sulphides (Carroll and Rutherford, 1988; Wallace and Carmichael, 1994; Pingitore, Meitzner and Love, 1997). The L spectrum of Fe also shows significant valence-related effects, but these are difficult to interpret unambiguously, owing to variable self-absorption and the influence of neighbouring atoms. A universally applicable microprobe method for Fe valency determination therefore appears unattainable, though empirical determination for specific minerals is more feasible. Methods applicable to glasses have been described by Matthews, Moncrieff and Carroll (1999) and Fialin *et al.* (2004).

7.9.2 *Mineral formulae*

The numbers of atoms derived from weight-percentage concentrations obtained by EMPA can be related to the relevant mineral formula (as in Table 7.3). For silicates it is usual to normalise with respect to an appropriate number of O atoms for the mineral concerned (see Table 7.4). (This number refers to oxygen associated with cations and excludes OH and H_2O, if present.) Also given in Table 7.4 are the theoretical numbers of cations. Agreement with the cation sum calculated from the analysis is a good test of the quality of the data.

More detailed structural formula calculations, in which assumptions about the distribution of cations between lattice sites are made, can be carried out. Programs have been developed by Richard and Clarke (1990) for amphiboles, and by Knowles (1987) for garnets. A spreadsheet-based program for amphiboles, which is convenient for processing large batches of tabulated data, has been described by Tindle and Webb (1994).

Unambiguous formula calculation for a mineral is not always possible, as in the case of micas containing lithium, though Tindle and Webb (1990) have shown that, in the case of trioctahedral micas (excluding those with high MgO content), an empirical relationship between Li_2O and SiO_2 contents can be used to estimate Li_2O contents for micas analysed with the electron microprobe.

7.9.3 *Data presentation*

The data output from quantitative EMPA can be presented in the form of tables of mass concentrations, or weight percentages, usually as oxides. It is normal practice to include a column of totals, which give an indication of analytical quality (in the case of anhydrous phases), or the presence of water. The number of significant figures used should not be excessive, to avoid giving a false impression of accuracy.

Presenting large numbers of analyses in tabular form is not only impracticable but also ineffective as a means of conveying the significance of the results. In some cases it is appropriate to give mean values for sets of analyses, with statistical data indicating the amount of scatter. However, it is often more relevant to use graphical forms of presentation revealing relationships between particular variables (Fig. 7.13). The variables used may be simple oxide weight-percentage concentrations, or derived quantities such as atom per cent, sums of elements occupying particular locations in structural formulae, molecular proportions of end-members, etc.

7.10 Standards

Pure elements may be used as standards for quantitative analysis but commonly are unsuitable for various reasons. Some elements, such as Cl, do not exist in solid form under normal conditions, while others are prone to oxidation in air or are difficult to polish. Also, using a pure element may sometimes result in excessive matrix corrections. Further, in WD analysis there may be enough difference in the peak wavelength between pure element and specimen, owing to chemical bonding effects, to cause significant errors.

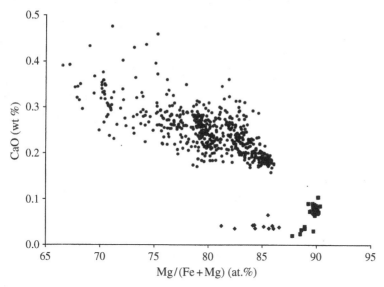

Fig. 7.13. Electron-microprobe data for volcanic olivines: circles, magma phenocrysts; diamonds, mantle xenocrysts; and squares, mantle xenoliths. (By courtesy of J. Johnson.)

Alternatives to pure elements standards are synthetic compounds and natural minerals, the former having the advantage of assured purity. Homogeneity on a micrometre scale is required. Also, immunity to oxidation or hydration in air and stability under vacuum are highly desirable. Some standards that are reasonably satisfactory according to the above criteria are listed in Table 7.5 (which is not exhaustive). These are available as raw materials from chemical suppliers (but note that most of the chemicals listed are normally in powder form, which is not ideal) or, in the case of minerals, from mineral collections or dealers. Also, mounted and polished standard sets can be obtained from specialist suppliers.

Complex minerals should be considered as 'reference samples' rather than standards used for primary calibration purposes, since the concentrations of individual elements may be relatively low. They can be analysed by EMPA using standards such as those described above to test whether satisfactory results are obtained for a mineral of the type concerned. Conventional chemical analysis requires the availability of an adequate amount of material, which should be homogeneous and inclusion-free. The testing and analysis of a range of minerals for use as reference samples has been described by Jarosewich, Nelen and Norberg (1979, 1980). Upper-mantle rocks are a valuable source of homogeneous monomineralic megacrysts, for example kaersutite from Kakanui, New Zealand (Reay, Johnstone and Kawachi, 1989).

Table 7.5. *Standards for quantitative electron microprobe analysis*

Element	Z	Pure element	Synthetic compounds	Natural minerals
Na	11		NaCl	Halite, albite, jadeite
Mg	12	*	MgO	Periclase, forsterite
Al	13	*	Al_2O_3	Corundum, kyanite
Si	14	*	SiO_2	Quartz, silicates
P	15		GaP	Apatite
S	16			Pyrite, anhydrite
Cl	17		NaCl	Halite
K	19		KCl	Orthoclase
Ca	20		CaF_2	Wollastonite, calcite
Sc	21	*		
Ti	22	*	TiO_2	Rutile, ilmenite
V	23	*	V_2O_5	
Cr	24	*	Cr_2O_3	Rhodonite, chromite
Mn	25	*		
Fe	26	*		Haematite, fayalite, pyrite
Co	27	*	CoO	
Ni	28	*	Ni_2Si	Millerite
Cu	29	*		Cuprite, chalcocite
Zn	30	*	ZnS	Willemite, sphalerite
Ga	31		GaP, GaAs	
Ge	32	*	GeO_2	
As	33	*	GaAs	Arsenopyrite
Se	34	*	$NbSe_2$	
Br	35	*	KBr	
Rb	37		RbCl	
Sr	38		$SrTiO_3$	Celestine
Y	39	*	$Y_3Al_5O_{12}$	Xenotime
Zr	40		ZrO_2	Zircon
Nb	41	*	Nb_2O_5, $Li_2Nb_2O_6$	Columbite
Mo	42	*	$CaMoO_4$	Molybdenite
Ru	44	*		
Rh	45	*		
Pd	46	*	PdTe	
Ag	47	*	Ag_2Te, Ag_2S	
Cd	48	*	CdS, CdSe, CdTe	
In	49	*	InP, InAs	
Sn	50	*	SnO_2, SnTe	Cassiterite
Sb	51	*		Stibnite
Te	52	*	TeO_2, In_2Te_3	
I	53		KI, CsI	
Cs	55		CsI	Pollucite
Ba	56		BaF_3	Barite, benitoite
La	57		LaB_6, LaF_3	
Ce	58		CeO_2, $CeAl_2$	
Pr	59		$PrAl_2$, $PrSi_2$, PrF_3	
Nd	60		$NdAl_2$, $NdSi_2$, NdF_3	

Table 7.5. (*cont.*)

Element	Z	Pure element	Synthetic compounds	Natural minerals
Sm	62		$SmAl_2$, SmF_2	
Eu	63		Eu_2O_3, EuF_3	
Gd	64	*	$GdAl_2$, GdF_3	
Tb	65	*	$TbAl_2$, $TbSi_2$, TbF_3	
Dy	66	*	$DyAl_2$, DyF_3	
Ho	67	*	$HoAl_2$, HoF_3	
Er	68	*	ErF_3	
Tm	69	*	$TmSi_2$, TmF_3	
Yb	70	*	YbF_3	
Lu	71	*	$LuSi_2$, LuF_3	
Hf	72	*	HfO_2	
Ta	73	*	Ta_2O_5	
W	74	*	$CaWO_4$	Wolframite
Rh	75	*		
Os	76	*		
Ir	77	*		
Pt	78	*		
Au	79	*		
Hg	80		HgTe	Cinnabar
Tl	81	*	TlI	Carlinite
Pb	82	*	PbO	Galena
Bi	83	*	Bi_2Te_3, Bi_2Se_3	
Th	90	*	ThF_4	Thorite
U	92	*	UO_2	Uraninite

*reasonably stable as pure element.

   Synthetic glasses are useful when natural minerals are not readily available, but can be produced only within certain compositional ranges. Also, they are prone to instability under electron bombardment, especially those containing alkalies; therefore it is desirable to use a defocussed beam and a low current for calibration measurements. Several glass standards have been developed by the US National Institute for Standards and Technology (Marinenko, 1991). Jarosewich and Boatner (1991) used a flux to grow crystals of rare-earth orthophosphates suitable for use as standards. McGuire, Francis and Dyar (1992) reported oxygen data obtained by neutron-activation analysis of various minerals intended for use as standards for O (avoiding dependence on an assumed stoichiometry).

   For quantitative ED analysis a profile including all the X-ray lines of a given element is required for spectrum fitting, and the relatively wide energy range involved must be devoid of peaks of other elements, whereas the requirement for WD analysis is merely that there should be no interferences at peak and background positions, which is more easily satisfied. Sometimes unwanted peaks can be 'stripped' from the ED spectrum of a standard, otherwise the use of a non-ideal standard such as a pure element may be unavoidable. In this case the effective peak height can be adjusted on the basis of an analysis of some other reference sample that has more lines in the spectrum and therefore is unsuitable for direct use.

7.10.1 Standardless analysis

In 'standardless' analysis calculated pure element intensities derived from theoretical or empirical data contained in the software are used rather than measured values (Labar, 1995). This option is offered in most ED systems. Its accuracy is difficult to evaluate, but certainly not as good as when 'real' standards are used.

Standardless WD analysis is more problematic, owing to the greater difficulty in predicting spectrometer efficiency. However, Fournier *et al.* (1999) have described a procedure that uses semi-empirical efficiency data to convert the intensities of peaks, as recorded in wavelength scans, into approximate concentrations.

8

X-ray analysis (2)

8.1 Light-element analysis

For present purposes 'light' elements are defined as those below 10 in atomic number. Of these, H and He do not produce characteristic X-rays, and the Li K line is outside the accessible wavelength range: the elements (and atomic numbers) concerned are thus Be (4), B (5), C (6), N (7), O (8) and F (9). The K lines of these elements are listed in Table 8.1. Light-element analysis requires an approach that differs in certain respects from that described in the previous chapter for 'ordinary' ($Z > 10$) elements.

The F Kα peak lies easily within the range of the TAP crystal, but O Kα is close to (for some instruments beyond) the upper limit of θ. Lead stearate ($2d = 100\,\text{Å}$) covers atomic numbers 5 (B) to 8 (O), but has been superseded by evaporated multilayers (Section 5.3.1), which give more intensity but poorer resolution. To detect long-wavelength X-rays a proportional counter with a thin window must be used (Section 5.3.4). There are potential interferences from high-order lines and the L and M lines of heavier elements, but these are small. Light-element K lines can be detected by thin-window ED detectors (Section 5.2.1) and are reasonably well resolved from each other.

Contamination with carbon (Section 3.10.1) gives rise to a spurious C peak, as well as causing additional absorption of emerging low-energy X-rays (especially O Kα, which is just above the absorption edge of carbon). Anti-contamination measures should therefore be used routinely for light-element analysis. In addition, specimen and holder should be thoroughly cleaned before introducing them into the instrument (see Section 9.7). The effect of carbon coating applied to insulating specimens can be avoided by using alternative coating materials (e.g. Al or Cu).

Table 8.1. *Data for K spectra of light elements*

Element	Atomic number	Excitation energy (eV)	Kα energy (eV)	Kα wavelength (Å)
Be	4	112	109	114.0
B	5	192	183	67.6
C	6	284	277	44.7
N	7	400	392	31.6
O	8	532	525	23.6
F	9	687	677	18.3

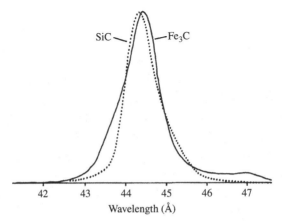

Fig. 8.1. Chemical effect on line shape: C K peaks recorded from SiC and Fe_3C, showing differences in position and shape. After Bastin and Heijligers (1986).

8.1.1 *Chemical bonding effects*

Light-element K spectra (produced by electron transitions between the L and K shells) are more dependent on chemical state than those of heavier elements, since the L shell is incomplete and the relevant energy levels are influenced by chemical bonding. In WD analysis this may require different spectrometer settings for specimen and standard. Further, where there are variations in the intensity distribution within the peak profile (Fig. 8.1), peak *area* is the preferred measure of intensity. Having once recorded and integrated the profile for a sample of a particular type, the relationship between peak height and area can be expressed as a 'profile factor' (or 'area/peak factor'), which may be used subsequently to convert peak heights into areas for any similar material.

8.1.2 Absorption corrections for light elements

For light elements, the absorption factor $f(\chi)$ may be much less than 1, since only a small fraction of the generated X-rays escape. Lowering the accelerating voltage decreases the depth of penetration of the electrons, and reduces the amount of absorption, so for light elements an accelerating voltage of 10 kV or less is advantageous. When the absorption correction is large, the shape of the depth distribution of X-ray production, $\phi(\rho z)$, is crucial, and the inadequacy of the classical ZAF model in the surface region becomes serious: 'phi–rho–z' models should therefore be used (Section 7.7.2). It is also particularly important that the specimen should be smooth and flat.

Mass absorption coefficients are problematic because there is a lack of data in the long-wavelength region, and the interpolation formulae used for 'normal' wavelengths do not apply. Special tables have been produced (e.g. Henke, Gullikson and Davis, 1993), but significant uncertainties still exist, especially in the vicinity of the L and M edges of heavier elements.

8.1.3 Application of multilayers

The F Kα line is much more intense with a multilayer than with a TAP crystal, though there is more overlap from neighbouring peaks such as Fe Lα and second-order Mg Kα (Potts and Tindle, 1989). The O Kα intensity is also much higher, which is advantageous for quantitative oxygen analysis (Nash, 1992). Nitrogen gives relatively poor intensity, but can be determined in feldspars containing ammonium ions (Beran, Armstrong and Rossman, 1992). An important application of multilayers is to boron, for which high intensities are obtainable (McGee and Anovitz, 1996). There is a strong interference from Cl Lα, for which a correction is necessary. Analysis of coal for carbon, oxygen and nitrogen has been described by Bustin, Mastalertz and Wilks (1993).

Multilayers perform most effectively at medium Bragg angles: at low angles the background intensity is high, whereas at high angles peak intensities are low (with no compensating improvement in resolution). Excess low-angle background is caused by specular reflection, which is especially marked for certain elements, including B, Si and Zr, which have X-ray lines of very long wavelength that are reflected (Rehbach and Karduck, 1992). Pulse-height analysis can be used to reduce this type of background.

8.2 Low-voltage analysis

As noted above, in light-element analysis it is desirable to use a low accelerating voltage in the interest of minimising absorption corrections. An incidental effect is that better spatial resolution is obtainable, owing to the decrease in electron range (Section 2.2.1), provided that the beam diameter is also reduced. The loss of current entailed with a conventional tungsten source can be overcome by using a field emission source (Section 3.2.1).

Enhanced resolution can be achieved for heavier elements too, provided that X-ray lines of suitably low excitation energy are used. This approach is known as low-energy X-ray emission spectroscopy (LEXES). With an accelerating voltage of 5 kV the K line is available only for atomic numbers below about 20, above which L lines must be used, and M lines must be used for atomic numbers above about 42. This entails a sacrifice in efficiency of X-ray generation. Also, the low-energy lines involved are subject to chemical bonding effects, and quantitative accuracy is less than when conventional conditions are used.

8.3 Choice of conditions for quantitative analysis

For quantitative analysis the accelerating voltage should preferably be at least twice the highest relevant excitation potential, in order to obtain reasonable intensities: 15 kV is thus the minimum when the heaviest element present is Fe, for instance. Peak intensities and peak-to-background ratios increase with increasing accelerating voltage (except that, in cases in which there is severe absorption of the emerging X-rays, the intensity decreases beyond a certain voltage). It follows that, by operating at a high voltage, better statistical precision can be obtained for a given beam current and counting time (or the same precision in a shorter time) and detection limits are lower. These advantages are offset, however, by the increased electron range in the sample, resulting in worsened spatial resolution and increased absorption corrections. Balancing these considerations leads to a choice within the range 10–25 kV in most cases.

A high beam current gives high X-ray intensities, but contrary factors should also be taken into account. For instance, samples prone to damage under electron bombardment may require the use of a low beam current (Section 8.7). Another consideration is that the dead-time of the X-ray counting system should not be excessive. For WD spectrometers a current in the range 10–100 nA is usual, except for trace elements requiring a higher current. For ED spectrometers the appropriate current is typically only a few nanoamps (unless the sensitivity has been reduced by placing an aperture in front of the detector).

It is difficult even for an experienced operator to arrive at optimal choices for all the relevant parameters in a multi-element WD analysis. A computer program that enables WD spectra to be simulated facilitates optimum choices of X-ray line, crystal, background offsets, etc. (Reed and Buckley, 1996) and is especially valuable in complex cases such as REE analysis (Reed and Buckley, 1998). The 'expert system' described by Fournier *et al.* (2000) considers all possible combinations of X-ray lines and spectrometer crystals. Count-rates are predicted by means of an X-ray-generation model combined with empirical spectrometer-efficiency data, and counting times chosen to give the statistical precision specified by the user. When interferences are significant, an alternative line (e.g. β instead of α, or L instead of K) is substituted. This procedure is repeated for a range of different accelerating voltages and the optimum config-uration determined automatically using the criterion of minimum total time per analysis. Some prior knowledge of at least the approximate composition is very desirable; if this information is not already available, it can be obtained by means of rapid 'standardless' analysis based on complete wavelength scans (see Section 7.10.1).

8.4 Counting statistics

X-ray photons are emitted randomly and intensities measured by counting pulses are therefore subject to statistical fluctuations. Repeated measurements of the number of counts, n, recorded in time t conform to a Gaussian distribu-tion, the width of which is dependent on the standard deviation, given by $\sigma = n^{0.5}$. The probability is 68% that a single measurement of n will lie within $\pm\sigma$ of the true value, 95% for $\pm 2\sigma$ and 99% for $\pm 3\sigma$. The uncertainty in a measured intensity may be expressed as the relative standard error, given by $\varepsilon = \sigma/n = n^{-0.5}$. Precision, defined as $\pm 2\varepsilon$, is plotted against n in Fig. 8.2. There is usually little to be gained by aiming for a precision of much better than $\pm 1\%$, since other factors limit ultimate accuracy: hence it is sufficient to accumulate approximately 10^5 counts.

The intensity of a peak, I_P, is given by the count-rate, which is equal to n_P/t_P, where n_P is the number of counts recorded in time t_P. The background intensity, I_B, is given by n_B/t_B, where n_B is the number of counts recorded in time t_B with the spectrometer offset to a background position (or the total number of counts measured on both sides of the peak). The standard deviations of peak and background count-rates are $n_P^{0.5}/t_P$ and $n_B^{0.5}/t_B$, respectively. The standard deviation (σ_{P-B}) of the background-corrected peak intensity ($I_P - I_B$) is given by

$$\sigma_{P-B} = [(n_P/t_P^2) + (n_B/t_B^2)]^{0.5}. \tag{8.1}$$

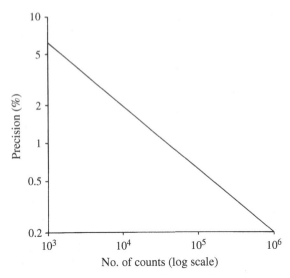

Fig. 8.2. Counting precision (two standard deviations) versus number of counts.

For example, suppose that $n_P = 54\,270$, $t_P = 10$, $n_B = 188$ and $t_B = 2$, then $I_P - I_B = 5427 - 94 = 5333$ counts per second and $\sigma_{P-B} = [(54\,270/100) + (188/4)]^{0.5} = 24$ counts per second. The relative precision of $I_P - I_B$ (defined as twice the relative standard error) is therefore $\pm 0.9\%$. For low concentrations the minimum total counting time $t_P + t_B$ for a given statistical error is obtained when $t_P/t_B = (I_P/I_B)^{0.5}$. At very low concentrations, for which $I_P \rightarrow I_B$, the optimum condition is $t_B = t_P$.

8.4.1 Homogeneity

It is sometimes useful to test a sample for homogeneity by determining the scatter in analyses of random points. A useful working criterion for homogeneity is that the scatter is less than three times the standard deviation as given by the square root of the number of counts. This assumes that other random errors are negligible, which is normally true.

Potts, Tindle and Isaacs (1983) described the application of a 'homogeneity index' (K) given by $K = \sigma C^{-0.5}$, where σ is the standard deviation of the concentration of a given element (the mean value of which is C) observed in a series of analyses. If K exceeds a certain value it may be concluded that the sample is inhomogeneous: the limiting K value can be derived from counting statistics or from empirical observations that include the effect of other fluctuations.

In general, natural variations in the concentrations of different elements in a given mineral are not independent: for example, in a ferromagnesian mineral an increase in Fe is normally accompanied by a decrease in Mg. This is taken into consideration in the treatment of Mohr, Fritz and Eckert (1990), in which observations are compared with computer-generated model populations in order to distinguish real inhomogeneity from random analytical fluctuations.

8.5 Detection limits

The 'detection limit' for a given element is the concentration which corresponds to a peak that can just be distinguished from statistical background fluctuations. A convenient working definition is that the peak height is equal to three standard deviations of the background count (the probability of this occurring by chance being less than 1%). This is illustrated in the following example: if the corrected pure-element count-rate is 10^4 counts per second, and the peak-to-background ratio is 10^3, then for a counting time of 10 s the detection limit expressed in counts is $3 \times (10^4 \times 10)^{0.5} = 30$, corresponding to a concentration of $(30/10^5) \times 100\% = 0.03\%$, which is a typical value for WD analysis of silicates, for instance (compared with about 0.1% for ED analysis).

By using conditions chosen specifically for trace-element analysis, a reduction of approximately a factor of ten in detection limit is attainable (Robinson, Ware and Smith, 1998; Reed, 2000). Statistical errors can be reduced by using a high beam current and long counting times for both peak and background (preferably dividing the counting times into alternating peak and background segments to minimise the effect of drift). Also, a higher than normal accelerating voltage enhances peak intensities and peak-to-background ratios. It is important to make allowances for background slope and curvature (Section 7.5.1) and interfering peaks (Section 7.5.2), in order to obtain reliable trace-element data. Since ideal conditions for trace and major elements differ, it is desirable to employ a separate procedure for each, with different accelerating voltages and beam currents.

8.6 The effect of the conductive coating

The conductive coating used on non-conductors (Section 9.5) affects X-ray intensities in two ways: firstly, incident electrons lose energy on passing through the coating, reducing the intensity of X-rays generated in the sample; and secondly, emerging X-rays are attenuated on passing through the carbon layer. The relative intensity loss $\Delta I / I$ can be derived from the following expression, which is valid for small values of the coating thickness, t:

$$\Delta I/I = \{[8.3 \times 10^5/(E_0^2 - E_c^2)] + \mu \operatorname{cosec} \psi\}\rho t, \qquad (8.2)$$

where E_0 is the incident electron energy, E_c is the excitation energy of the X-ray line concerned (both in keV), μ is the absorption coefficient of the coating material for the relevant X-ray wavelength, ψ is the X-ray take-off angle and ρ is the density of the coating. A density value of $2.0 \, \text{g cm}^{-3}$ for evaporated carbon has been reported (Jurek, Renner and Krousky, 1994). The first term in Eq. (8.2) represents the effect of electron energy loss. For Fe Kα radiation the intensity loss due to this factor is 0.6% ($E_0 = 20$ keV), but rises to 2.2% at 12 keV. The second term accounts for X-ray absorption. The effect of the carbon coating on peak intensities is generally small and can be minimised by coating standards (even if they are conductors) as well as specimens and taking steps to ensure that the thickness is constant (see Section 9.5.1).

8.7 Beam damage

Although the beam current used in microprobe analysis is generally low, its effects are concentrated in a small volume of the sample and are not always negligible, necessitating preventive measures in some cases, as described in the following sections.

8.7.1 Heating

Most of the energy in the electron beam is converted into heat at the point of impact. Though the power dissipated is only of the order of milliwatts, the power density is quite high. The heating effect is small in the case of a metal with high thermal conductivity, but a material with low thermal conductivity may experience a considerable local rise in temperature (see Section 2.9), which can be enough to cause damage, such as loss of carbon dioxide from carbonates and water from hydrous minerals. The loss of such components causes the apparent concentrations of the remaining elements to increase. In WD analysis, in which X-ray lines are measured sequentially, progressive degradation of the sample during analysis will affect lines measured later more than those measured earlier, therefore incorrect element ratios may be obtained.

Heating can be reduced by applying a coating of a good conductor (e.g. aluminium, copper, or silver) of thickness greater than necessary merely for electrical conduction (Smith, 1986). (The choice of coating element is influenced by whether its X-ray lines are a nuisance.) A correction for the loss of X-ray intensity can be calculated, or else obtained from an empirical working curve relating the measured intensity of the coating element to the factor by which the peak intensity of the analysed element is reduced (see Fig. 8.3).

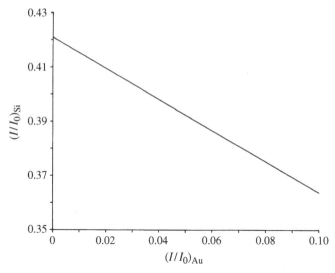

Fig. 8.3. The ratio $(I/I_0)_{Si}$ of Si Kα intensity measured on Au-coated SiO_2 relative to pure Si, with various thicknesses of Au, plotted against the ratio $(I/I_0)_{Au}$ of Au Mα intensity measured on SiO_2 relative to pure Au. (Accelerating voltage 10 kV). This plot can be used to correct for the effect of the Au coating. After Willich and Obertop (1990).

8.7.2 Migration of alkalies etc.

A well-known example of the effect of ion migration under the influence of the electrostatic field produced by the electron beam is the decrease in Na Kα intensity with time as Na^+ ions move away from the bombarded area, which occurs in glasses, feldspars and some other phases. Similar, though less marked, behaviour is shown by potassium. Under some conditions the apparent alkali concentration may increase above its true value after an initial decrease. Preventive measures include reducing the beam current or defocussing the beam, the latter generally being more effective (see Fig. 8.4). Scanning the beam in a raster, or moving the specimen continuously during the analysis, are other possibilities. Also, alkali migration can be inhibited by cooling with liquid nitrogen, though most instruments lack the necessary cooling arrangements.

ED analysis has a considerable advantage over WD analysis, in that adequate X-ray intensities can be obtained with a lower beam current. Increasing the accelerating voltage reduces the rate of alkali migration (Goodhew and Gulley, 1975). A procedure for extrapolating the variation of Na Kα intensity back to zero time in order to estimate the 'true' initial intensity

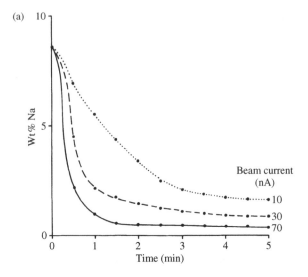

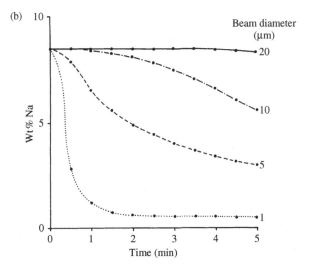

Fig. 8.4. Loss of Na from albite as a function of time (accelerating voltage 15 kV): (a) beam diameter 1 μm, various currents; and (b) beam current 50 nA, various diameters. After Autefage and Couderc (1980).

has been described by Nielsen and Sigurdsson (1981). Techniques for hydrous glasses have been investigated by Morgan and London (1996, 2005).

Stormer, Pierson and Tacker (1993) observed variations in the intensity of the F Kα emission from apatite depending on crystal orientation. This behaviour is explained by anisotropic diffusion of F ions to the surface, followed by loss of F to the vacuum over a longer time scale. Somewhat similar, though less extreme, behaviour is shown by Cl. Loss of P from phosphates can occur at high beam currents (Jercinovic and Williams, 2005). Other minerals that exhibit effects due to electron bombardment include realgar (AsS) and cinnabar (HgS).

8.8 Boundary effects

The results of quantitative analyses close to boundaries between phases are affected by the finite size of the X-ray source. Assuming that the beam diameter is small ($<1\,\mu m$), the source size (and hence the spatial resolution) is governed by the penetration of the electrons and the spreading of their trajectories in the specimen. The effective spatial resolution in quantitative analysis can be defined as the size of the region within which 99% of the measured characteristic X-rays are produced, and is related to the electron range (Section 2.2.1). The *effective* range in the present context is the distance travelled before the energy falls to the critical excitation energy, E_c, beyond which no more characteristic X-rays are produced. The diameter d (in micrometres) of the source region can be estimated from the following equation:

$$d = 0.22(E_0^{1.5} - E_c^{1.5})/\rho, \tag{8.3}$$

where E_0 and E_c are in keV and ρ is the density of the specimen. For example, in the case of a silicate with a density of $3\,g\,cm^{-3}$ and with $E_0 = 15\,keV$ and $E_c = 4\,keV$ (Ca), $d = 3.7\,\mu m$.

Finite spatial resolution modifies element profiles when the rate of change is significant on a micrometre scale, for example in zoned minerals and in artificial diffusion profiles used to determine diffusion coefficients. The latter case has been analysed by Ganguly, Bhattacharya and Chakrabarty (1988), who proposed a method of 'deconvolution' based on the assumption of a Gaussian X-ray source profile, which enables the original profile to be retrieved.

Fluorescence effects at boundaries can be significant and may extend over significant distances (see Section 7.7.3).

8.9 Special cases

Quantitative electron microprobe analysis is normally carried out on flat well-polished specimens using a focussed electron beam at normal incidence. If these conditions are not satisfied the accuracy of the results may suffer. However, the cases of analysis under non-ideal conditions described in the following sections are of practical interest and steps can be taken to minimise the loss of accuracy.

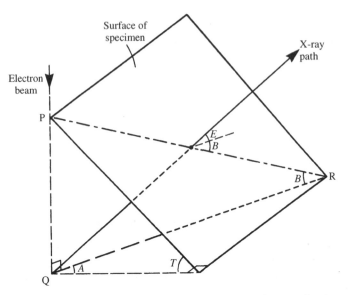

Fig. 8.5. The geometry of absorption correction for a tilted specimen (see text).

8.9.1 Tilted specimens

In an SEM the specimen may be tilted, which changes both the X-ray take-off angle and the angle of electron incidence. Assuming that the shape of the $\phi(\rho z)$ function remains the same, the effect of tilt can be corrected by replacing cosec ψ in the absorption correction by $\cos T / (\sin T \cos A \cos E + \sin E \cos T)$, where T (tilt angle), A (azimuth angle) and E (elevation angle) are as shown in Fig. 8.5. (The shape of $\phi(\rho z)$ is affected somewhat by the angle of electron incidence, but this is usually neglected, as is the effect of tilt on the backscattering correction).

If the specimen surface has irregular topography, the tilt angle is not known. A procedure for dealing with such cases described by Wiens *et al.* (1994) involves rotating the specimen through 180° and repeating the X-ray intensity measurements on the same point. Reasonable results can be obtained by simply averaging these values, provided that the tilt angle is not too large.

8.9.2 Broad-beam analysis

An essential characteristic of EMPA is its spatial resolution (normally approximately 1 μm), but sometimes it is appropriate to use a deliberately broadened beam (e.g. to determine the average composition of a devitrified melt). The beam can be enlarged for this purpose by defocussing the final lens, or alternatively the beam may be scanned in a raster. When WD spectrometers are used, the size of the analysed area should be limited to less than 100 µm to minimise spectrometer defocussing, which affects different elements to a varying degree. Larger areas can be analysed by EDS (the count-rate should be restricted, however, in order to avoid differential dead-time effects between areas of varying composition when scanning is used).

The weighted sum of the matrix effects for the individual phases within the analysed area is not the same as that of an equivalent homogeneous sample, because of the nonlinear dependence of correction factors on concentration. The results obtained by applying matrix corrections to broad-beam analyses are therefore not as accurate as for normal point analyses of single phases.

8.9.3 Particles

Small particles (less than about 25 µm in diameter) are difficult to mount and polish. They can, however, be analysed in an unpolished state, mounted on an SEM stub, for example. Particles of the order of 1 µm in size should preferably be supported on a thin carbon film, which gives minimal X-ray background. Very small particles (less than 1 µm in diameter) should be analysed in an analytical electron microscope (Section 1.4.1).

The usual procedure for quantitative analysis is liable to give inaccurate results for particles, since the correction formulae are based on the ideal of a flat well-polished specimen. Also, if the particle is too small to absorb the entirety of the beam, the X-ray intensities are reduced by comparison with solid standards (Fig. 8.6). The problem of loss of X-ray generation can be avoided by deriving concentration *ratios* from the ratios of X-ray line intensities: for example, Pyman, Hillyer and Posner (1978) obtained linear calibration plots of concentration ratios versus intensity ratios for clay mineral particles in the 1–5-µm size range. However, this does not make any allowance for differential absorption effects between X-ray lines of different elements.

A more rigorous approach is to use modified corrections that take account of the particle geometry. In the Armstrong–Buseck method (Armstrong, 1991), correction equations derived for various idealised particle geometries are used, the nearest to the actual geometry being applied in each case. Particle shapes and sizes are assessed by means of SEM images or optical microscopy. Alternatively, the Monte Carlo method (Section 2.8) can be used to simulate X-ray generation in particles of various geometries.

Another possibility is to measure peak-to-background ratios and make use of the fact that the effect of particle geometry on the continuum is similar to that on characteristic X-rays of the same energy (Statham and Pawley, 1977; Small, Newbury and Myklebust, 1979). Concentrations can be derived from

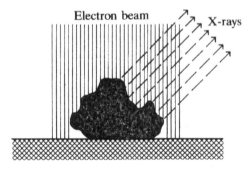

Fig. 8.6. Particle analysis: results are affected by loss of X-ray intensity because some electrons in the beam miss the particle (when this is smaller than the beam diameter) and the X-ray absorption path varies owing to the irregular shape.

peak-to-background ratios measured on the sample compared with ratios measured on standards. In ED spectra it is often necessary to remove the peaks by 'stripping' in order to determine the background, owing to the lack of suitable peak-free regions in the spectrum. The precision of measured peak-to-background ratios is governed by the statistical error in the relatively low background intensity: this necessitates longer acquisition times than are customarily used for measuring peaks.

8.9.4 Rough and porous specimens

Absorption and other corrections are affected by roughness (irregularities of much less than 1 μm can have a significant effect). The angle of the surface is also important and should not vary by more than 1°. Care should therefore be exercised in order to avoid polishing relief at the edges of grains.

Porosity also affects the results of quantitative analysis. If the pores are filled with mounting medium, a fraction of the incident electron energy will be absorbed in the medium, causing a loss of X-ray production in the sample itself and consequently low analytical results. If the pores are vacant the effect is less marked. Also, extraneous material introduced during polishing can lead to misleading results (e.g. spurious Al due to alumina).

The method based on peak-to-background ratios described in the preceding section can be applied to rough and porous samples, in order to obtain at least semi-quantitative results.

8.9.5 Thin specimens

Limits to spatial resolution governed by the penetration and spreading of the beam in thick specimens can be overcome by using thin (e.g. 100 nm) specimens through which electrons pass with relatively little scattering. Rock and mineral specimens prepared as for transmission electron microscopy (usually by ion-beam thinning) can be used. Either an electron microprobe or SEM with an ED spectrometer can be utilised for the analysis of such samples, but it is better to use an analytical electron microscope (AEM), as described in Section 1.4.1, with which spatial resolution of about 10 nm is possible.

ED analysis is used in this field because of the need for high detection efficiency to compensate for the low X-ray intensity. Methods of spectrum processing used for thick specimens are applicable (see Section 7.6). Somewhat different considerations apply, however, to the conversion of peak intensities into concentrations, because the intensities depend on specimen thickness, which is difficult to determine. The usual approach is to derive concentration ratios from relative peak intensities within the spectrum. This requires information on the 'sensitivity factor' (intensity per unit concentration) for each element. This can be deduced from the efficiency of X-ray generation, which may be calculated from first principles, and the detection efficiency, which can also be calculated given knowledge of the thickness of the detector window etc. Alternatively a purely empirical calibration curve based on measurements on specimens of known composition can be used (Cliff and Lorimer, 1975). For sufficiently thin specimens the effects of absorption, fluorescence, etc. can be neglected, but in practice an absorption correction is often required for geological samples containing elements such as Na and Mg. For further details, see Joy, Romig and Goldstein (1986); and for a discussion of geological aspects, see Champness (1995).

8.9.6 Fluid inclusions

The analysis of fluid inclusions is a challenging problem, but residues left after opening fluid-containing cavities by cleaving or fracturing the host crystals can be analysed (Eadington, 1974), as can crystals that coexist with the fluid (Metzger *et al.*, 1977; Anthony, Reynolds and Beane, 1974). The results are, however, only semi-quantitative, owing to the effects of topographic irregularity. Aqueous fluids can be analysed *in situ* if frozen. Initial freezing should be rapid in order to minimise segregation of ice; the samples must then be kept cold while the host crystals are cleaved or fractured, given a conducting coating, and transferred to a cold stage in the instrument. Ayora and Fontarnau (1990) used standards consisting of frozen solutions containing different known amounts of Na, K and Ca chlorides, for quantitative analysis of frozen natural fluids of similar composition. The beam current must be low in order to avoid damage to the sample.

8.9.7 *Analysis in low vacuum*

Scanning electron microscopy in the 'low-vacuum' or 'environmental' mode, which is useful for samples that are hydrous and/or uncoated (Section 3.10.2), can be combined with X-ray analysis, with certain provisos. The gas pressure in the sample chamber is too low for there to be any significant effect on the detection of the X-rays from the specimen, but additional peaks (e.g. O K from water molecules) may appear owing to the interaction of electrons with gas atoms. More importantly, scattering by gas atoms causes an extensive 'skirt' to exist around the electron beam, which can excite X-rays from surrounding objects, thereby degrading the effective spatial resolution. Various measures can be applied to minimise this effect, including using a high accelerating voltage, a gas of low atomic number, low gas pressure and minimum electron path length in the gas. Various methods for correcting X-ray analyses for the residual effect remaining even when such steps are taken have been proposed (Newbury, 2002).

9

Sample preparation

9.1 Initial preparation of samples

Most geological specimens require some preliminary treatment before mounting for examination in the SEM or analysis by EMPA. Often cleaning is necessary in order to eliminate unwanted contaminants. Sediments (and soils) commonly need drying. Friable and porous materials usually require impregnation, especially if polished samples are to be produced. Hand specimens have to be cut to a slice of an appropriate size for mounting and polishing. These processes are described in the following sections.

For further information on these and other aspects of specimen preparation the reader is referred to Humphries (1992), Laflamme (1990), Miller (1988), and Smart and Tovey (1982).

9.1.1 Cleaning

As collected, many samples contain unwanted components, which hinder examination of the specific features of interest and need to be removed. For example, sediments and soils often require washing with distilled water to remove soluble salts (mainly chlorides). Only gentle agitation should be used as a rule, ultrasonic cleaning being liable to damage the mineral grains. Unwanted carbonate can be removed with hydrochloric acid, iron oxides with stannous chloride and organic matter with potassium permanganate or hydrogen peroxide. Hydrocarbons can be removed by soaking in a solvent such as trichloroethane (pressure may be required in the case of low-porosity materials).

9.1.2 Drying

Some sample materials are wet in their normal state and must be dried. This can be carried out by gentle heating in air (temperatures above about 50 °C can

149

cause loss of structural water from clay minerals). Other approaches are required, however, when fragile structures need to be preserved. Damage can be limited by replacing water with a volatile liquid of lower surface tension, such as amyl acetate, before drying. Other techniques, developed for drying fragile biological materials, can be adapted for clays and soils (McHardy, Wilson and Tait, 1982). For example, in 'freeze drying' water is removed from the frozen sample by sublimation in vacuum. The sample must be frozen rapidly to minimise growth of ice crystals: this is achieved by immersing the sample (which should be as small as possible) in a liquid cooled by liquid nitrogen.

The least damaging, but slowest, technique is 'critical-point drying', which relies on conversion of liquid into vapour above the temperature of the critical point, so that there is no phase change. For water this temperature is inconveniently high, so the water in the sample is replaced by a more suitable liquid before drying. A typical procedure entails replacing the water first by methanol, then by liquid carbon dioxide. The sample temperature is raised just above the critical point of the latter ($32\,°C$) and carbon dioxide is vented slowly from the chamber.

9.1.3 Impregnation

Friable materials require impregnation with a suitable medium such as a low-viscosity epoxy resin to impart the necessary mechanical strength for normal specimen-preparation procedures to be applied. Also, the filling of pores and cavities is desirable in order to avoid entrapment of polishing materials etc. and outgassing in the instrument vacuum. In some cases it is desirable to dilute the medium with a solvent such as toluene or acetone.

The effectiveness of impregnation can be improved by removing air in a vacuum chamber and then applying atmospheric pressure to force the medium into the pores. This can be achieved by lowering the sample into a liquid medium under vacuum, then admitting air. (For epoxy resins the pressure should not be below 10 torr, or excessive frothing will occur.) Alternatively, the liquid medium is poured onto the sample (Fig. 9.1) and, on admitting air into the sample chamber, is forced into the pores. To achieve maximum penetration, several pumping and venting cycles may be necessary. Higher pressures can be applied by using gas from a cylinder. Gamberini and Valdrè (1995) have described a preparation procedure for pumice, involving repeated vacuum embedding and grinding.

9.1.4 Replicas and casts

For applications in which pore structures are of interest, it is useful to produce replicas or casts for examination in the SEM. This entails impregnation, as

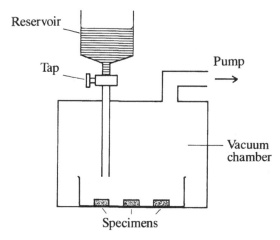

Fig. 9.1. Vacuum impregnation: the chamber is evacuated to remove air from pores in specimens; liquid embedding medium (e.g. epoxy resin) is introduced by opening a stopcock; on venting the chamber, medium is forced into pores by air pressure.

described above, followed by dissolution of the sample material, using hydrochloric acid for carbonates and hydrofluoric acid for silicates. Repeated pumping and venting cycles can be used to obtain maximum penetration by the medium. Details of methods suitable for chalk specimens have been given by Walker (1978) and Patsoules and Cripps (1983).

Latex rubber casts of fossil plant impressions can be used for scanning electron microscopy (Chaloner and Gay, 1973). Latex may be applied in several successive layers, allowing each to dry before applying the next. The resulting cast is adequately resistant to vacuum and electron bombardment for SEM examination.

9.1.5 Cutting rock samples

Hand specimens of rocks require cutting to give pieces of suitable size and shape for mounting and polishing. Usually cutting is done with a circular diamond saw, a parallel-sided slice (typically a few millimetres thick) being cut off and then trimmed to the size required for the mounted section. Friable specimens need to be impregnated (as described in Section 9.1.3) before cutting. Damage to the specimen occurring during these operations may extend to considerable depth, possibly affecting the final product; therefore the methods used should be as gentle as possible.

9.2 Mounting

9.2.1 The SEM 'stub'

Specimens for SEM examination are commonly mounted on a 'stub', which takes the form of a disc, usually made of aluminium and typically about 1 cm in diameter, with a spigot for attachment to the stage mechanism (Fig. 3.10). (Sometimes graphite stubs are used, to minimise X-ray background, especially for particulates.) The sample is glued to the stub and coated to provide conduction, as described in Section 9.5. A quick alternative method of attachment is to use double-sided sticky tape, though this is to be avoided if possible. Quick-setting glue can also be used when speed is important. Mounting materials and adhesives should have a low vapour pressure so that the instrument vacuum is not adversely affected. Whatever arrangement is used, there must be an electrical path to the holder: if necessary this should be provided by applying carbon or silver paint (Fig. 9.2). For small specimens such as microfossils the use of a vacuum-compatible wax has some advantages (Finch, 1974). The wax is warmed so that it flows over the surface of the stub and the specimens are pressed into the wax while it is soft. For some purposes it is desirable to mount an electron-microscope grid with numbered bars on the stub, so that individual specimens can be relocated by means of their 'grid reference'. The special problems arising in the mounting of soil samples have been treated in detail by Lohnes and Demirel (1978).

9.2.2 Embedding

For some specimens (e.g. ore minerals) it is not necessary to use thin sections; instead, embedding in a solid block is appropriate. The sample is placed in a mould made of non-stick material, and the embedding medium in liquid form poured in (Fig. 9.3). Alternatively, a metal or plastic ring can be temporarily

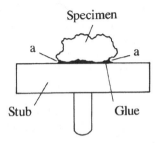

Fig. 9.2. An SEM specimen mounted on a 'stub'; note that conductive coating may not connect across overhanging regions (a), which should therefore be painted with silver or carbon 'paint'.

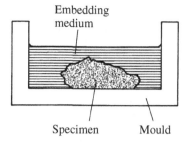

Embedding
medium

Specimen   Mould

Fig. 9.3. Embedding a specimen for polishing: liquid medium (e.g. epoxy resin) is poured into a 'non-stick' mould and allowed to set (with heating if necessary).

stuck down to serve the same purpose, remaining part of the mount after the medium has set. Bakelite (supplied as a powder and polymerised by application of pressure and heat) can be used for embedding, but epoxy resins that are either cold setting or require only relatively mild heating are less likely to cause damage to the specimen. Bubbles can be removed by applying moderate vacuum. There is some advantage in using a conducting medium, such as epoxy resin filled with fine carbon or metal particles.

9.2.3 Thin sections

In many applications, thin sections are required so that viewing by transmitted light is possible. Preparation is as for ordinary thin sections in the first stages, the rock slice being attached to the glass slide using one of the special epoxy resins with suitable optical properties that are available. The slice is ground to a thickness somewhat greater than the 30 μm final thickness required, before commencing polishing. Long microscope slides can be shortened by cutting the ends off, in order to fit the specimen holder more conveniently. Also sometimes used, especially in the USA, are 1-inch (25.4-mm)-diameter round sections.

9.2.4 Grain mounts

Special techniques are required for mounting small grains. One possibility is to mix the grains with embedding medium and set in a mould. Alternatively, they may be pressed into a thin layer of epoxy resin on a glass cover slip. After setting, this is inverted and mounted on a glass slide, and the cover slip ground away, leaving the grains lying in one plane ready for polishing (Fig. 9.4). To avoid clumps, mixing with crushed graphite particles of similar size prior to embedding is an effective way of keeping the grains separate.

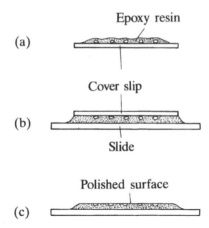

Fig. 9.4. Mounting small grains: (a) grains embedded in a thin layer of epoxy resin on a cover slip; (b) embedded grains mounted on a microscope slide; and (c) cover slip ground away and exposed grains polished.

For SEM work, polishing might not be required. In this case, grains can be scattered onto a sticky surface or partly dried carbon paint, etc., or alternatively suspended in a liquid, drops of which are transferred to the substrate and then evaporated (in the latter case, it may be necessary to de-coagulate the particles with the aid of an ultrasonic bath). Suitable substrate materials include beryllium (which gives minimal X-ray emission), carbon (which is almost as good in this respect, but is difficult to obtain with a very smooth surface) and silicon (which can be highly polished, but emits more X-rays, which is a drawback in some applications.)

9.2.5 Standards

Methods used for mounting standards are essentially similar to those already described. Usually fairly small (e.g. a few millimetres) pieces of standard materials are used, so a number can be mounted in the specimen holder at one time. Standards can be prepared individually, allowing selection of those required for each application, or alternatively a single block containing many standards can be used. This saves space, but obtaining a good polish on a wide variety of materials simultaneously is difficult. Prepared standard blocks are obtainable from commercial suppliers. It is convenient for commonly used standards to be mounted in the specimen holder semi-permanently. Those needed only occasionally for specific applications can be loaded temporarily for calibration and then removed to make space for specimens.

9.3 Polishing

For X-ray analysis, mapping and BSE imaging it is extremely desirable to avoid topographic effects: specimens therefore should be flat and well polished. Polishing procedures for ore microscopy can be adapted to rocks consisting predominantly of silicates. Starting with a flat ground surface, polishing is carried out with progressively finer grades of abrasive (typically carborundum or emery for the coarser grades and diamond or alumina in the later stages). Paper or woven nylon laps are preferable to cloth with a 'nap', since they have less tendency to produce surface relief between minerals of different hardnesses. Either rotating or vibrating motion of the lap is used, the former being preferable. The specimens should be thoroughly cleaned after each stage, to avoid transfer of abrasive material in pores and cracks. For soft phases a final hand polish using very fine alumina may be necessary. A single-stage polishing technique using only alumina has been described by Allen (1984). Special procedures are required for electron backscatter diffraction studies (Section 4.8.3), the damaged surface layer left by conventional polishing being removed by a final polish with alkaline colloidal silica slurry (Lloyd *et al.*, 1981).

Polished thin sections are usually made by polishing a section that has previously been ground to a thickness somewhat greater than 30 μm (Fig. 9.5). However, the final thickness is then poorly constrained, which is undesirable for polarised-light microscopy. This can be avoided by polishing one face of the rock slice first, temporarily mounting it face down on a glass slide, grinding off surplus material to give a thickness of 30 μm, mounting permanently with epoxy resin on another glass slide (ground face down), and finally removing the first slide. After polishing, specimens should be cleaned by washing in a solvent that does not attack the mounting medium (e.g. ethanol or petroleum ether), preferably using an ultrasonic bath to dislodge remnants of polishing materials. When surface contamination is especially important (e.g. for light-element analysis) plasma cleaning is desirable (Isabell *et al.*, 1999).

9.4 Etching

Chemical etching enables chemical and crystallographic differences to be converted into topography that can be observed in secondary-electron images. (It is inappropriate for quantitative EMPA, for which flat, smooth, surfaces are required, and etching may alter surface composition.) Carbonates can be etched with dilute hydrochloric acid (1–5%), acetic acid (20%), or EDTA, the most delicate effect being obtained with the last two. In some cases heavy etching to remove carbonate cement, leaving exposed grains of quartz etc. for SEM study,

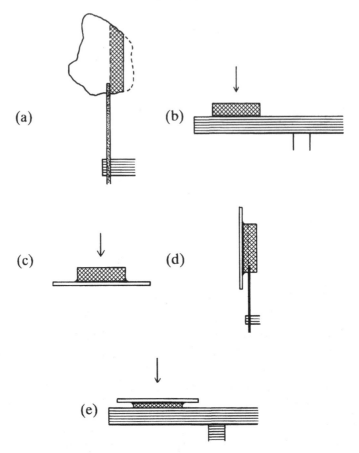

Fig. 9.5. Preparation of a polished thin section: (a) a slice of rock is cut with a diamond saw and trimmed to size; (b) one face of the slice is ground and lapped; (c) the slice is attached to a glass slide and lapped face down; (d) surplus material is cut off with a fine diamond saw; and (e) the surface is ground and polished leaving ∼30 μm thickness.

is appropriate. Quartz grains may be etched with concentrated hydrochloric acid for studies of surface texture. Etching of polished sections with hydrofluoric acid can be used to reveal textures of quartzitic sandstones, for example, and fine exsolution textures in silicates (Section 4.5.3). The sections can be suspended above an acid bath, to be etched by the fumes, or immersed for a stronger effect. (The glass slide may be protected by covering it with paraffin wax.)

9.5 Coating

Most geological samples, being non-conductors of electricity, require a conductive coating to prevent charging under electron bombardment unless this is

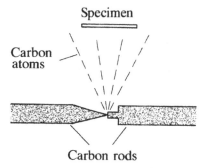

Fig. 9.6. Evaporation of carbon to give conductive coating on sample, using pointed carbon rods and large current.

avoided by selecting a low accelerating voltage (Section 4.6.2) or by using an 'environmental' or 'low-vacuum' SEM (Section 3.10.2). The preferred coating element for X-ray analysis is carbon, because it has a minimal effect on the X-ray spectrum. It is also the best choice for cathodoluminescence studies. However, it is not ideal for SEM imaging, owing to its low secondary-electron yield. For this purpose a metal such as gold is preferable, or alternatives with finer grain structure, such as gold–palladium alloy, chromium or iridium, but these are less suitable for X-ray analysis and BSE imaging. Gold coating is also advantageous for fibrous clay minerals, which tend to collapse under the electron beam when coated with carbon (Purvis, 1991).

9.5.1 Carbon coating

The usual method of coating with carbon is to place the specimen in a vacuum chamber with a carbon evaporation source consisting of pointed carbon rods (3–6 mm in diameter) in contact (Fig. 9.6). A current of about 100 A is passed through the rods for a few seconds, causing carbon to be evaporated from the region where the rods are in contact. The pressure should be less than approximately 10^{-4} torr, as obtained with either a diffusion pump or a turbo pump. (Carbon films produced under poor vacuum conditions are 'sooty' and lack adhesion.) Since the evaporated carbon atoms travel in straight lines, this coating method is suitable only for flat specimens, not for irregularly shaped objects, better coverage of which can be obtained by rotating the sample during coating.

The optimum thickness of carbon is about 20 nm. The thickness can be controlled approximately by using a fixed current and evaporation time. It can also be estimated from the colour of the coated surface of a polished metal such as brass (Kerrick, Eminhizer and Villaume, 1973): orange corresponds to

15 nm, indigo red to 20 nm, blue to 25 nm and bluish green to 30 nm. More accurate monitoring can be achieved by means of a quartz crystal forming part of an electronic oscillator circuit, with one surface exposed to the evaporant, the oscillator frequency being used to indicate coating thickness. Other methods of thickness monitoring are based on the electrical resistance or optical density of the coating. Specimens should be equi-distant from the carbon source in order to ensure that uniformity of coating thickness is attained.

9.5.2 Metal evaporation

Although carbon is usually the preferred coating material for samples that are to be analysed, the enhanced thermal conductivity obtained with a metal coat is advantageous in reducing the effects of electron bombardment for certain types of sample (Section 8.7). Gold, as used for SEM samples, is best avoided because it interacts strongly with electrons and X-rays on account of its high atomic number. Alternatives such as aluminium, copper and silver are therefore sometimes used. Coating with these metals can be carried out by vacuum evaporation, using a wire basket made of tungsten, or a 'boat' made from molybdenum sheet, heated by passing a large electric current (Fig. 9.7). In some cases sputter coating may be used instead (see the next section).

9.5.3 Sputter coating

This is a convenient method for producing coatings of metals, including gold and gold–palladium alloys. Since a relatively poor vacuum is used, sputtered atoms are strongly scattered by gas molecules and travel in all directions,

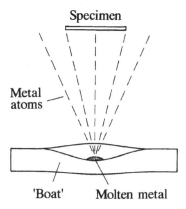

Fig. 9.7. Evaporation of metal (e.g. Ag, Al) by heating in a molybdenum 'boat' carrying a large current.

which is advantageous for coating specimens of irregular shape. The 'diode' type of sputter coater is shown in Fig. 9.8. The chamber is first evacuated with a rotary pump; argon is then admitted to give a pressure of about 10^{-1} torr. When a high voltage is applied across the electrodes a discharge occurs and the metal foil target is bombarded with ions, which remove atoms by sputtering. These are deposited onto the specimen, the coating thickness being determined by the discharge current and time. The current is controlled by varying the argon pressure. The specimen is heated significantly by electron bombardment, which can cause damage to fragile materials. 'Cold' sputter coaters, in which the electrons are deflected by a magnet, are available. For high-resolution SEM work, metal films of better quality can be obtained using a turbo-pumped sputter coater.

9.5.4 Removing coatings

It may be desirable to remove a coating applied for SEM or EMP work, for example to allow unhindered examination by optical microscopy, or to permit a gold coating applied for SEM examination to be replaced by carbon for X-ray analysis (or vice versa). Carbon deposited under high-vacuum conditions adheres strongly to the substrate, but can be removed from polished sections with a fine polishing medium (e.g. 0.25-μm diamond on a cloth lap). Gold and other metals can easily be wiped off polished specimens. Traces remaining in

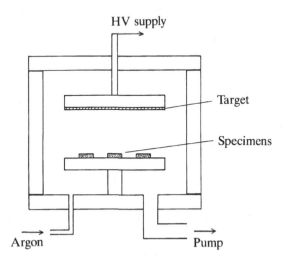

Fig. 9.8. Sputter coating: air is removed from the chamber and replaced by argon at low pressure; high voltage (HV) applied to the top electrode causes a discharge in the gas; specimens are coated with metal atoms (e.g. Au) removed from the target by 'sputtering' due to bombardment with argon ions.

cracks, etc., may be removed with potassium iodide solution. Specimens with strong three-dimensional topography require a different approach. Gold can be removed from such samples by treatment with a 10% aqueous solution of sodium cyanide (Sela and Boyde, 1977), with appropriate precautions in view of its toxicity. Silver has the advantage that it can be removed easily with photographic 'Farmer's reducer' (Mills, 1988).

9.6 Marking specimens

Specimens must, of course, be marked for identification purposes. Aluminium SEM stubs can be inscribed with a fine metal stylus, or alternatively written on with a pen. In the case of blocks of epoxy resin in which opaque specimens are mounted for polishing, a label can be embedded with the specimen, or else the identification number can be written or scratched onto the back. For thin sections a number may be inscribed on the back using a diamond point.

It is also sometimes desirable to mark areas of interest within specimens to make them easier to find in the electron microprobe or SEM. Ink rings can be drawn on the back of thin sections for this purpose, if the instrument concerned has transmitted-light viewing facilities. Rings drawn on the *front* are naturally appropriate for opaque specimens and when transmitted-light viewing is unavailable: for this purpose either conducting ink should be used or else the carbon coating must be applied after drawing the rings. Ink that is unaffected by the solvents used for cleaning should be selected. The need for marking can be obviated by using other approaches described in the following sections.

9.6.1 Specimen 'maps'

Finding areas of interest using the optical microscope in the electron microprobe can be difficult because of the high magnification and small field of view. Scanning electron images, which allow much lower magnification, can be used instead, but relevant features are sometimes not easily identifiable in such images. Time can therefore be saved by employing stratagems described below.

A sketch map or low-magnification photograph of the whole specimen is an invaluable aid. A 'macro' photograph can be used, but scanned digital images are now preferred. Micrographs of small areas at higher magnifications are also sometimes useful and can be taken with an ordinary photomicrography set-up. Opaque grains provide useful 'landmarks' in plain-light micrographs that can easily be recognised either in the built-in microscope (in the case of the electron microprobe) or in backscattered electron images.

The need for a 'map' may be obviated by using a bench microscope with x and y coordinate read-out. Full microscope facilities – low- and high-power objectives, polarised light, etc. – are available for finding areas of interest. The positions of these can be recorded relative to a reference mark, so that the same points may be found easily after transfer to the electron microprobe or SEM (provided that this has calibrated stage movements).

9.7 Specimen handling and storage

Specimens and standards should be kept in a dust-free environment, preferably in a desiccator. For pure-element standards that oxidise readily and minerals such as sulphides that tend to tarnish it is desirable to use a vacuum desiccator. SEM 'stubs' can be stored in plastic boxes available for the purpose.

Cleanliness should be observed when handling specimens, to avoid degradation of the specimen-chamber vacuum and enhancement of the rate of deposition of carbon contamination under the beam (gloves should preferably be used). Specimens can be cleaned by washing with a residue-free solvent and wiping with a tissue. The solvent should be one that does not attack the specimen, mounting medium, or ink used for labelling. Suitable choices include petroleum ether and ethanol. Dust can be removed with a compressed-air jet.

Appendix

The following figures show ED spectra of various minerals, as recorded with a thin-window Si(Li) detector. The abscissae show the energy in keV. The accelerating voltage is 20 kV and the X-ray take-off angle is 30°. Only major (α) peaks of elements are labelled; the C K peak from the carbon coating is not shown. (Note that relative peak intensities vary with accelerating voltage and, in the low-energy region, are affected by the window characteristics.) Many minerals exhibit a considerable range of compositions, which should be taken into account.

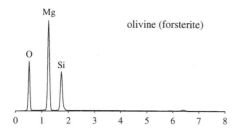

olivine (forsterite)

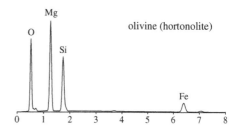

olivine (hortonolite)

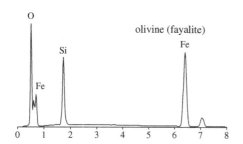

olivine (fayalite)

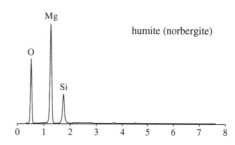

olivine (tephroite)

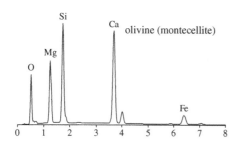

olivine (montecellite)

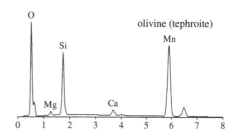

humite (norbergite)

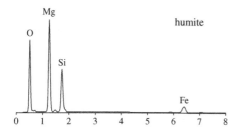

humite

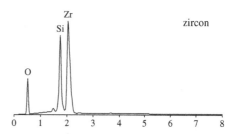

zircon

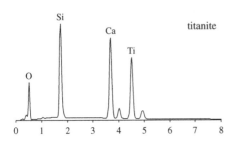

titanite

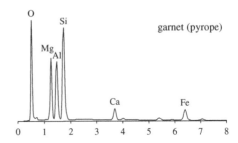

garnet (pyrope)

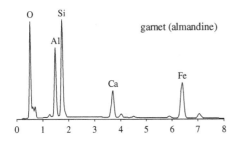

garnet (almandine)

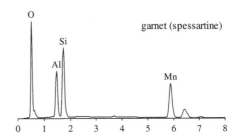

garnet (spessartine)

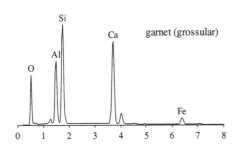

garnet (grossular)

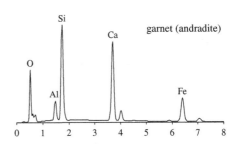

garnet (andradite)

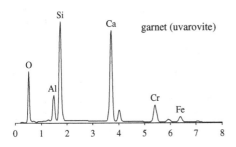

garnet (uvarovite)

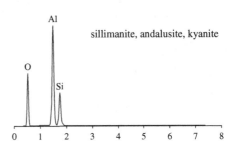

sillimanite, andalusite, kyanite

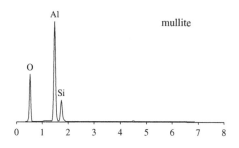

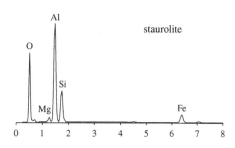

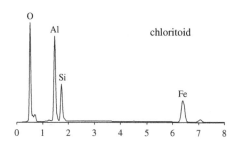

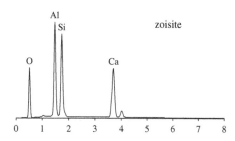

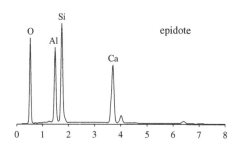

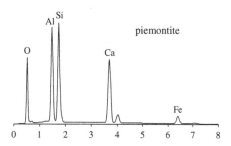

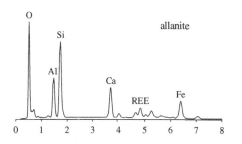

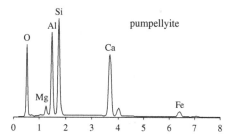

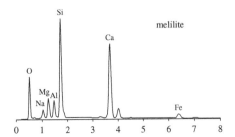

melilite

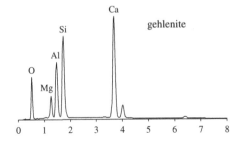

gehlenite

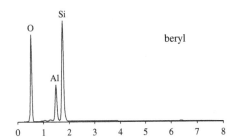

beryl

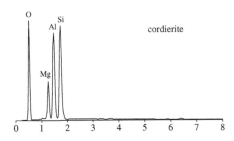

cordierite

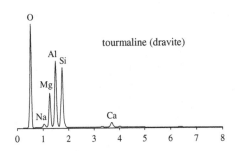

tourmaline (dravite)

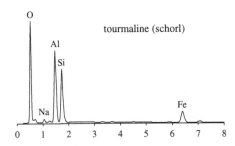

tourmaline (schorl)

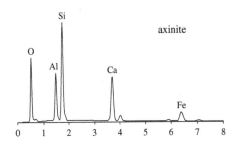

axinite

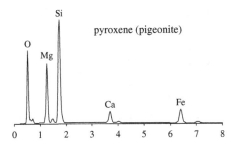

pyroxene (pigeonite)

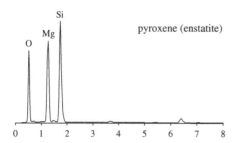

pyroxene (enstatite)

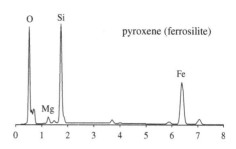

pyroxene (ferrosilite)

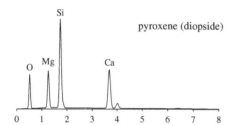

pyroxene (diopside)

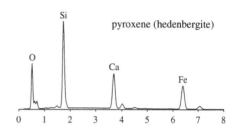

pyroxene (hedenbergite)

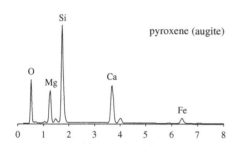

pyroxene (augite)

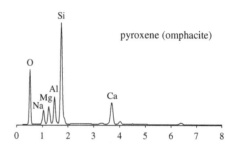

pyroxene (omphacite)

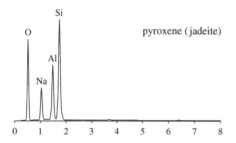

pyroxene (jadeite)

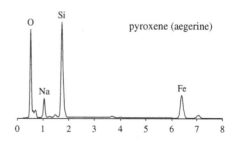

pyroxene (aegerine)

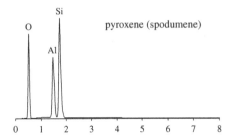

pyroxene (spodumene)

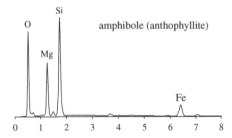

amphibole (anthophyllite)

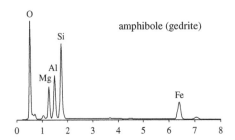

amphibole (gedrite)

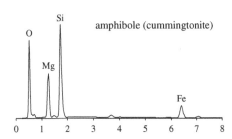

amphibole (cummingtonite)

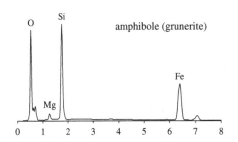

amphibole (grunerite)

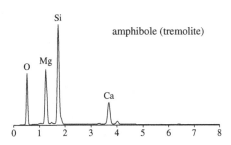

amphibole (tremolite)

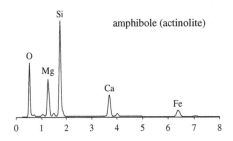

amphibole (actinolite)

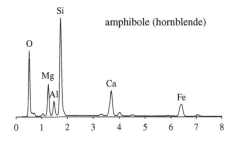

amphibole (hornblende)

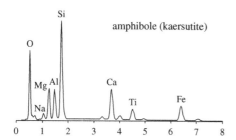

amphibole (kaersutite)

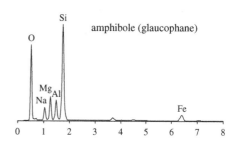

amphibole (glaucophane)

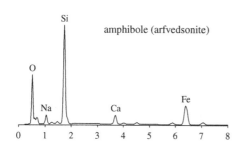

amphibole (riebeckite)

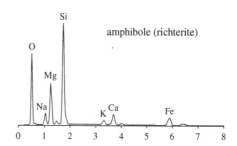

amphibole (richterite)

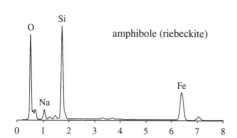

amphibole (arfvedsonite)

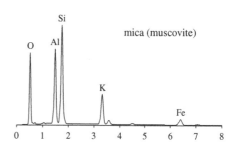

mica (muscovite)

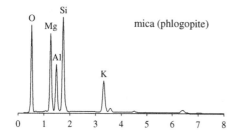

mica (phlogopite)

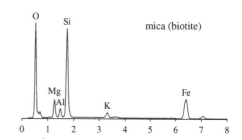

mica (biotite)

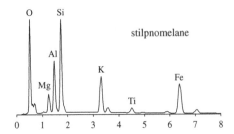

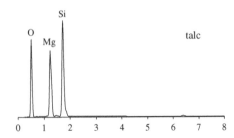

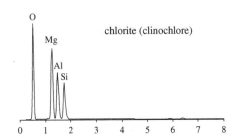

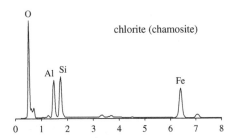

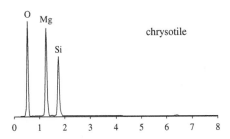

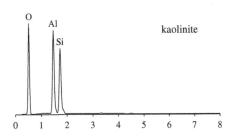

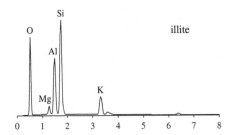

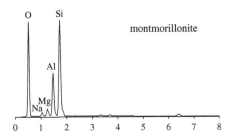

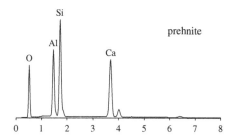

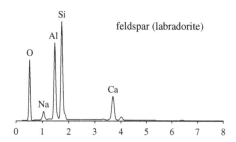

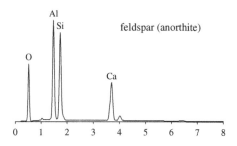

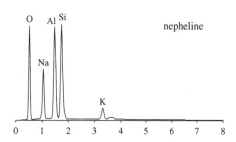

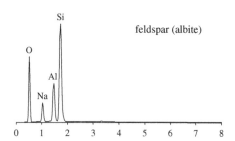

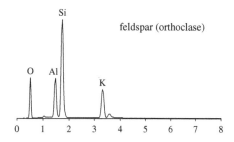

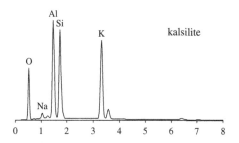

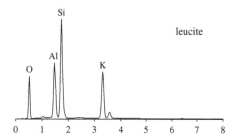

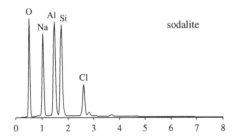

sodalite

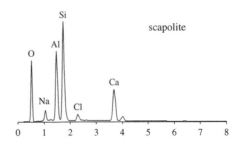

scapolite

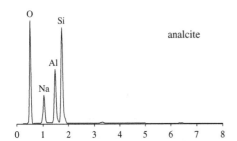

analcite

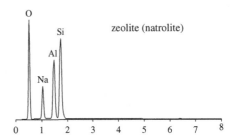

zeolite (natrolite)

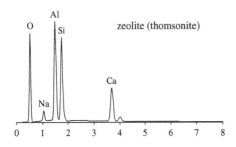

zeolite (thomsonite)

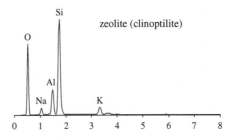

zeolite (clinoptilite)

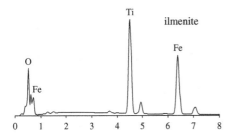

ilmenite

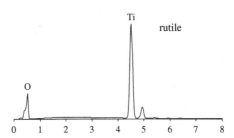

rutile

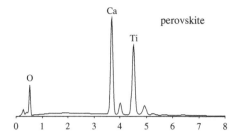

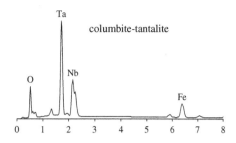

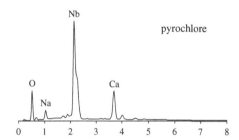

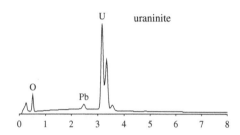

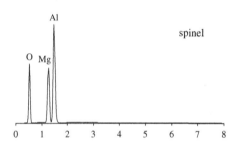

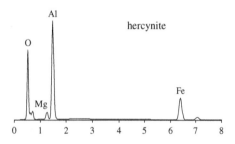

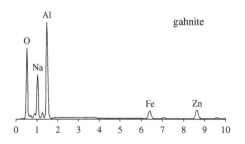

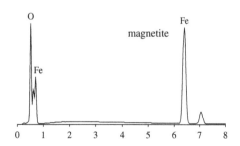

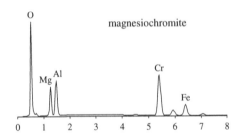

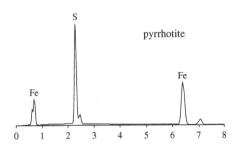

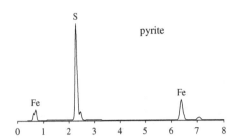

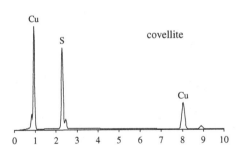

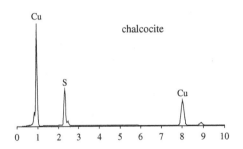

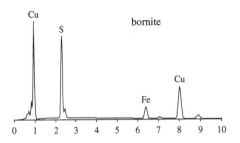

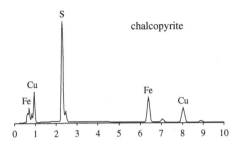

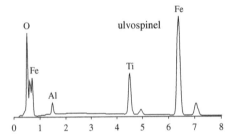

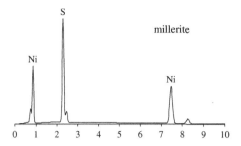

millerite

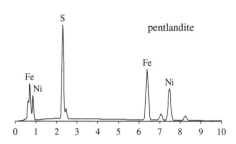

pentlandite

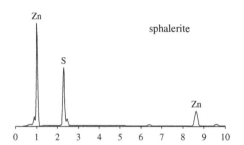

sphalerite

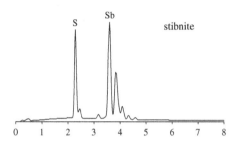

stibnite

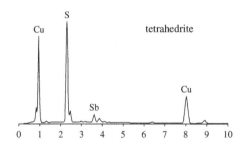

tetrahedrite

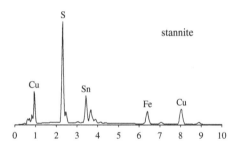

stannite

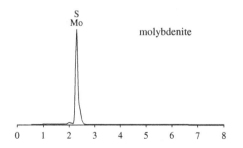

molybdenite

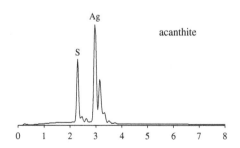

acanthite

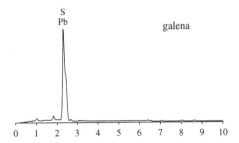

galena

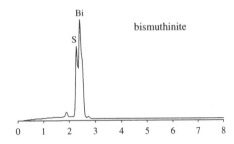

bismuthinite

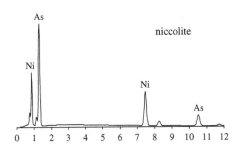

arsenopyrite

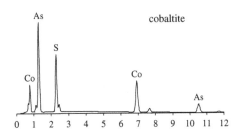

cobaltite

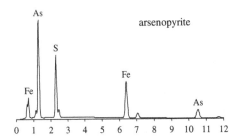

niccolite

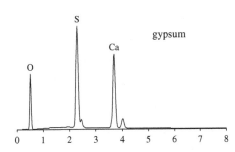

gypsum

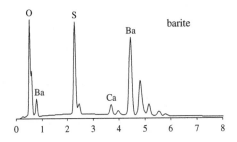

barite

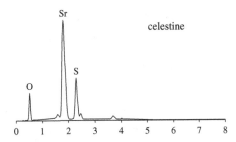

celestine

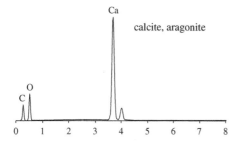

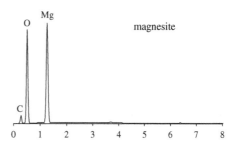

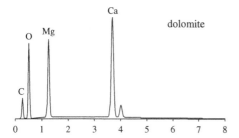

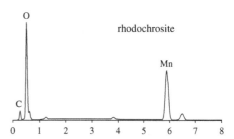

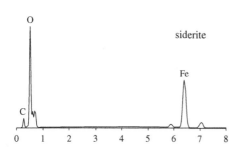

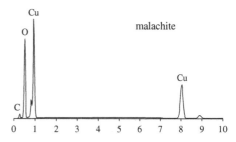

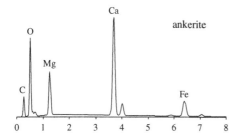

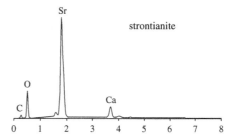

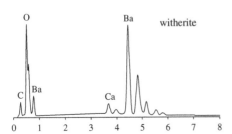

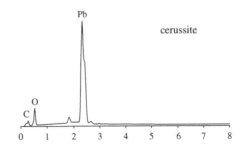

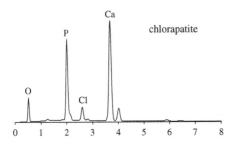

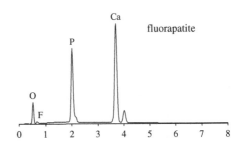

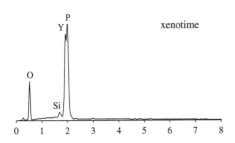

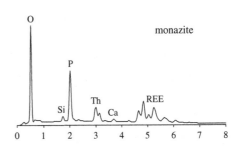

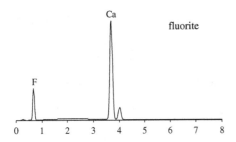

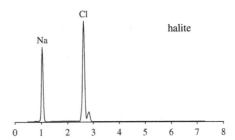

References

Albee, A. L. and Ray, L. (1970) Correction factors for electron probe microanalysis of silicates, oxides, carbonates, phosphates, and sulfates. *Anal. Chem.* **42** 1408–14.

Allen, D. (1984) A one-stage precision polishing technique for geological specimens. *Mineral. Mag.* **48** 298–300.

Anthony, E. Y., Reynolds, T. J. and Beane, R. E. (1974) Identification of daughter minerals in fluid inclusions using scanning electron microscopy and energy dispersive analysis. *Amer. Mineral.* **59** 1053–7.

Armstrong, J. T. (1991) Quantitative elemental analysis of individual microparticles with electron beam instruments. In *Electron Probe Quantitation*, ed. K. F. J. Heinrich and D. E. Newbury (New York: Plenum Press) pp. 261–315.

— (1995) CITZAF: a package of correction programs for the quantitative electron microbeam X-ray analysis of thick polished materials, thin films, and particles. *Microbeam Anal.* **4** 177–200.

Autefage, F. and Couderc, J.-J. (1980) Etude du mécanisme de la migration du sodium et du potassium au cours de leur analyse à la microsonde électronique. *Bull. Minéral.* **203** 623–9.

Ayora, C. and Fontarnau, R. (1990) X-ray microanalysis of frozen fluid inclusions. *Chem. Geol.* **89** 135–48.

Bastin, G. F. and Heijligers, H. J. M. (1986) Quantitative electron probe microanalysis of carbon in binary carbides. I – Principles and procedures. *X-Ray Spectrom.* **15** 135–41.

Bence, A. E. and Albee, A. L. (1968) Empirical correction factors for the electron microanalysis of silicates and oxides. *J. Geol.* **76** 382–403.

Beran, A., Armstrong, J. and Rossman, G. R. (1992) Infrared and electron microprobe analysis of ammonium ions in hyalophane feldspar. *Eur. J. Mineral.* **4** 847–50.

Boggs, S., Krinsley, D. H., Goles, G. G., Seyedolali, A. and Dypvik, H. (2001) Identification of shocked quartz by scanning cathodoluminescnce imaging. *Meteoritics Planet. Sci.* **36** 783–91.

Boyde, A. (1979) The perception and measurement of depth in the SEM. *Scanning Electron Microsc.* **1979/II** 67–78.

Bright, D. S. (1992) Visibility of two intermixed phases as a function of grain size and signal-to-noise: a computer simulation. In *Proc. 50th Annual Meeting of the Electron Microscopy Society of America*, ed. G. W. Bailey, J. Bentley and J. A. Small (San Francisco, CA: San Francisco Press) pp. 1610–11.

179

Bustin, R. M., Mastalertz, M. and Wilks, K. R. (1993) Direct determination of carbon, oxygen and nitrogen content in coal using the electron microprobe. *Fuel* **72** 181–5.

Cabri, L. J. and Campbell, J. L. (1998) The proton microprobe in ore mineralogy (micro-PIXE technique). In *Modern Approaches to Ore and Environmental Mineralogy*, ed. L. J. Cabri and D. J. Vaughan, Short Course Series, vol. 27 (Ottawa: Mineralogical Association of Canada) pp. 181–98.

Carroll, M. R. and Rutherford, M. J. (1988) Sulfur speciation in hydrous experimental glasses of varying oxidation state: results from measured wavelength shifts of sulfur X-rays. *Amer. Mineral.* **73** 845–9.

Chaloner, W. G. and Gay, M. M. (1973) Scanning electron microscopy of latex casts of fossil plant impressions. *Palaeontology* **16** 645–9.

Champness, P. E. (1995) Analytical electron microscopy. In *Microprobe Techniques in the Earth Sciences*, ed. P. J. Potts, J. F. W. Bowles, S. J. B. Reed and M. R. Cave (London: Chapman and Hall) pp. 91–139.

Chapman, P. A. and Meagher, E. P. (1975) A technique for observing exsolution lamellae in pyroxenes with the scanning electron microscope. *Amer. Mineral.* **60** 155–6.

Cliff, G. and Lorimer, G. W. (1975) The quantitative analysis of thin specimens. *J. Microsc.* **103** 203–7.

Cosca, M. A., Essene, E. J. and Bowman, J. R. (1991) Complete chemical analyses of metamorphic hornblendes: implications for normalizations, calculated H_2O activities, and thermobarometry. *Contrib. Mineral. Petrol.* **108** 472–84.

Dalton, J. A. and Lane, S. J. (1996) Electron microprobe analysis of Ca in olivine close to grain boundaries: the problem of secondary X-ray fluorescence. *Amer. Mineral.* **81** 194–201.

Danilatos, G. D. (1994) Environmental scanning electron microscopy and microanalysis. *Mikrochim. Acta* **114/5** 143–55.

Demars, C., Pagel, M., Deloule, E. and Blanc, P. (1996) Cathodoluminescence of quartz from sandstones: interpretation of the UV range by determination of trace element distributions and fluid inclusion $P–T–X$ properties in authigenic quartz. *Amer. Mineral.* **81** 891–901.

D'Lemos, R. S., Kearsley, A. T., Pembroke, J. W., Watt, G. R. and Wright, P. (1997) Complex quartz growth histories in granite revealed by scanning cathodoluminescence techniques. *Geol. Mag.* **134** 549–52.

Donovan, J. J., Snyder, D. A. and Rivers, M. L. (1993) An improved interference correction for trace element analysis. *Microbeam Anal.* **2** 23–8.

Droop, G. T. R. (1987) A general equation for estimating Fe^{3+} concentrations in ferromagnesian silicates and oxides from microprobe analyses, using stoichiometric criteria. *Mineral. Mag.* **51** 431–5.

Eadington, P. J. (1974) Microprobe analysis of the non-volatile components in fluid inclusions. *Neues Jahrb. Mineral., Monatsh.* 518–25.

Fagan, T. J., Taylor, G. J., Keil, K. *et al.* (2003) Northwest Africa 773: lunar origin and iron-enrichment trend. *Meteoritics Planet. Sci.* **38** 529–54.

Feenstra, A. and Engi, M. (1998) An experimental study of the Fe-Mg exchange between garnet and ilmenite. *Contrib. Mineral. Petrol.* **131** 379–92.

Fialin, M. (1988) Modification of Philibert–Tixier ZAF correction for geological samples. *X-ray Spectrom.* **17** 103–6.

(1992) Background determination in wavelength-dispersive electron microprobe analysis: some difficulties and presentation of a new analytical model. *X-ray Spectrom.* **21** 175–81.

Fialin, M., Bézos, A., Wagner, C., Magnien, V. and Humler, E. (2004) Quantitative electron microprobe analysis of $Fe^{3+}/\Sigma Fe$: basic concepts and experimental protocol for glasses. *Amer. Mineral.* **89** 654–62.

Finch, E. M. (1974) An improved method of mounting palaeontological specimens for scanning electron microscope examination. *Palaeontology* **17** 431–4.

Fournier, C., Merlet, C., Dugne, O. and Fialin, M. (1999) Standardless semi-quantitative analysis with WDS-EPMA. *J. Anal. Atom. Spectrom.* **14** 381–6.

Fournier, C., Merlet, C., Staub, P. F. and Dugne, O. (2000) An expert system for EPMA. *Mikrochim. Acta* **132** 531–9.

Fraser, D. G. (1995) The nuclear microprobe – PIXE, PIGE, RBS, NRA and ERDA. In *Microprobe Techniques in the Earth Sciences*, ed. P. J. Potts, J. F. W. Bowles, S. J. B. Reed and M. R. Cave (London: Chapman and Hall) pp. 141–62.

Gamberini, F. and Valdrè, G. (1995) Preparative method and analysis by OM, SEM and EPMA of porous, brittle and low permeability rocks and materials: the case of pumices. *Microsc. Microanal. Microstruct.* **6** 573–86.

Ganguly, J., Bhattacharya, R. M. and Chakrabarty, S. (1988) Convolution effect in the determination of compositional profiles and diffusion coefficients by microprobe step scans. *Amer. Mineral.* **73** 901–9.

Ginibre, C., Kronz, A. and Wörner, G. (2002) High-resolution quantitative imaging of plagioclase composition using accumulated backscattered electron images: new constraints on oscillatory zoning. *Contrib. Mineral. Petrol.* **142** 436–48.

Goldstein, J. I., Newbury, D. E., Joy, D. C. *et al.* (2003) *Scanning Electron Microscopy and X-Ray Microanalysis* (New York: Kluwer Academic/Plenum).

Goncalves, P., Williams, M. L. and Jercinovic, M. J. (2005) Electron-microprobe age mapping of monazite. *Amer. Mineral.* **90** 578–85.

Goodhew, P. J. and Gulley, J. E. C. (1975) The determination of alkali metals in glasses by electron microprobe analysis. *Glass Technol.* **15** 123–6.

Halden, N. M., Campbell, J. L. and Teesdale, W. J. (1995) PIXE analysis in mineralogy and geochemistry. *Canad. Mineral.* **33** 293–302.

Harris, D. H. (1990) Electron-microprobe analysis. In *Advanced Microscopic Studies of Ore Minerals*, ed. J. L. Jambor and D. J. Vaughan, Short Course Handbook no. 17 (Ottawa: Mineralogical Association of Canada) pp. 319–39.

Heinrich, K. F. J. and Newbury, D. E. (1991) *Electron Probe Quantitation* (New York: Plenum Press).

Henke, B. L., Gullikson, E. M. and Davis, J. C. (1993) X-ray interactions: photoabsorption, scattering, transmission, and reflection at $E = 50$–$30,000$ eV, $Z = 1$–92. *Atom. Data Nucl. Data Tables* **54** 181–342.

Herd, C. D. K., Papike, J. J. and Brearley, A. J. (2001). Oxygen fugacity of martian basalts from electron microprobe oxygen and TEM–EELS analysis of Fe–Ti oxides. *Amer. Mineral.* **86** 1015–24.

Higgins, S. J., Taylor, L. A., Chambers, J. G., Patchen, A. and McKay, D. S. (1996). X-ray digital-imaging petrography: technique development for lunar soils. *Meteoritics Planet. Sci.* **31** 356–61.

Hinton, R. W. (1995) Ion microprobe analysis in geology. In *Microprobe Techniques in the Earth Sciences*, ed. P. J. Potts, J. F. W. Bowles, S. J. B. Reed and M. R. Cave (London: Chapman and Hall) pp. 235–89.

Hochella, M. F. (1988) Auger electron and X-ray photoelectron spectroscopies. *Rev. Mineral.* **18** 573–637.

Hochella, M. F., Harris, D. W. and Turner, A. M. (1986) Scanning Auger microscopy as a high-resolution microprobe for geologic materials. *Amer. Mineral.* **71** 1247–57.

Humphries, D. W. (1992) *The Preparation of Thin Sections of Rocks, Minerals, and Ceramics* (Oxford: Oxford University Press).

Isabell, T. C., Fischione, P. E., O'Keefe, C., Guruz, M. U. and Dravid, V. P. (1999) Plasma cleaning and its applications for electron microscopy. *Microsc. Microanal.* **5** 126–35.

Jacobson, C. E. (1989) Estimation of Fe^{3+} from electron microprobe analyses: observations on calcic amphibole and chlorite. *J. Metamorphic Geol.* **7** 507–13.

Jarosewich, E. and Boatner, L. A. (1991) Rare-earth element reference samples for electron microprobe analysis. *Geostand. Newslett.* **15** 397–9.

Jarosewich, E., Nelen, J. A. and Norberg, J. A. (1979) Electron microprobe reference samples for mineral analyses. *Smithson. Contrib. Earth Sci.* **22** 68–72.

(1980) Reference samples for electron microprobe analysis. *Geostand. Newslett.* **4** 43–8.

Jercinovic, M. J. and Williams, M. L. (2005) Analytical perils (and progress) in electron microprobe trace element analysis applied to geochronology: Background acquisition, interferences, and beam irradiation effects. *Amer. Mineral.* **90** 526–46.

Joy, D. C. (1995) *Monte Carlo Modeling for Electron Microscopy and Microanalysis* (New York: Oxford University Press).

Joy, D. C., Romig, A. D. and Goldstein, J. I., eds. (1986) *Principles of Analytical Electron Microscopy* (New York: Plenum Press).

Jurek, K., Renner, O. and Krousky, E. (1994) The role of coating densities in X-ray microanalysis. *Mikrochim. Acta* **114/115** 323–6.

Kanaya, K. and Okayama, S. (1972) Penetration and energy-loss theory of electrons in solid targets. *J. Phys. D* **5** 43–58.

Kerrick, D. M., Eminhizer, L. B. and Villaume, J. F. (1973) The role of carbon film thickness in electron microprobe analysis. *Amer. Mineral.* **58** 920–5.

Kloprogge, J. T., Boström, T. E. and Weier, M. L. (2004) *In situ* observation of the thermal decomposition of weddelite by heating stage environmental scanning electron microscopy. *Amer. Mineral.* **89** 245–8.

Knowles, C. R. (1987) A BASIC program to recast garnet end members. *Computers Geosci.* **13** 655–8.

Krinsley, D. H., Pye, K., Boggs, S. and Tovey, N. K. (1998) *Backscattered Scanning Electron Microscopy and Image Analysis of Sediments and Sedimentary Rocks* (Cambridge: Cambridge University Press).

Labar, J. L. (1995) Standardless electron probe X-ray analysis of non-biological samples. *Microbeam Anal.* **4** 65–83.

Laflamme, J. H. G. (1990) The preparation of materials for microsocopic study. In *Advanced Microscopic Studies of Ore Minerals*, ed. J. L. Jambor and D. J. Vaughan (Ottawa: Mineralogical Association of Canada) pp. 37–68.

Lane, S. J. and Dalton, J. A. (1994) Electron microprobe analysis of geological carbonates. *Amer. Mineral.* **79** 745–9.

Lastra, R., Petruk, W. and Wilson, J. (1998). Image-analysis techniques and applications to mineral processing. In *Modern Approaches to Ore and Environmental Mineralogy*, ed. L. J. Cabri and D. J. Vaughan, Short Course Series, vol. 27 (Ottawa: Mineralogical Association of Canada) pp. 327–66.

Laubach, S. E., Reed, R. M., Olson, J. E., Lander, R. H. and Bonnell, L. M. (2004) Coevolution of crack–seal texture and fracture porosity in sedimentary rocks: cathodoluminescence observations of regional fractures. *J. Struct. Geol.* **26** 967–82.

Llovet, X. and Galan, G. (2003) Correction of secondary X-ray fluorescence near grain boundaries in electron microprobe analysis: application to thermobarometry of spinel lherzolites. *Amer. Mineral.* **88** 121–30.

Lloyd, G. E. (1987) Atomic number and crystallographic contrast images in the SEM: a review of backscattered electron techniques. *Mineral. Mag.* **51** 3–19.

Lloyd, G. E., Hall, M. G., Cockayne, B. and Jones, D. W. (1981) Selected-area channelling patterns from geological materials: specimen preparation, indexing and representation of patterns, and applications. *Canad. Mineral.* **19** 505–18.

Lohnes, R. A. and Demirel, T. (1978) SEM applications in soil mechanics. *Scanning Electron Microsc.* **1978/I** 643–54.

Maaskant, P. and Kaper, H. (1991) Fluorescence effects at phase boundaries: petrological implications for Fe–Ti oxides. *Mineral. Mag.* **55** 277–9.

Marinenko, R. B. (1991) Standards for electron probe microanalysis. In *Electron Probe Quantitation*, ed. K. F. J. Heinrich and D. E. Newbury (New York: Plenum Press) pp. 251–60.

Markowitz, A. and Milliken, K. L. (2003) Quantification of brittle deformation in burial compaction, Frio and Mount Simon Formation sandstones. *J. Sed. Res.* **73** 1007–21.

Marshall, D. L. (1988) *Cathodoluminescence of Geological Materials* (Boston: Unwin Hyman).

Matthews, S. J., Moncrieff, D. H. S. and Carroll, M. R. (1999) Empirical calibration of the sulphur valence oxygen barometer from natural and experimental glasses: method and applications. *Mineral. Mag.* **63** 421–31.

McGee, J. J. and Anovitz, L. M. (1996) Electron microprobe analysis of geologic materials for boron. In *Boron: Mineralogy, Petrology and Geochemistry*, ed. E. S. Grew and L. M. Anovitz, Reviews of Mineralogy, vol. 33 (Washington: Mineralogical Society of America) pp. 771–88.

McGuire, A. V., Francis, C. A. and Dyar, M. D. (1992) Mineral standards for electron microprobe analysis of oxygen. *Amer. Mineral.* **77** 1087–91.

McHardy, W. J., Wilson, M. J. and Tait, J. M. (1982) Electron microscope and X-ray diffraction studies of filamentous illitic clay from sandstones of the Magnus Field. *Clay Mineral.* **17** 23–39.

McMahon, G. and Cabri, L. J. (1998) The SIMS technique in ore mineralogy. In *Modern Approaches to Ore and Environmental Mineralogy*, ed. L. J. Cabri and D. J. Vaughan, Short Course Series, vol. 27 (Ottawa: Mineralogical Association of Canada) pp. 153–80.

Metzger, F. W., Kelly, W. C., Nesbitt, B. E. and Essene, E. J. (1977) Scanning electron microscopy of daughter minerals in fluid inclusions. *Econ. Geol.* **72** 141–52.

Miller, J. (1988) Microscopical techniques: 1. Slices, slides, stains and peels. In *Techniques in Sedimentology*, ed. M. Tucker (Oxford: Blackwells) pp. 86–107.

Mills, A. A. (1988) Silver as a removable conductive coating for scanning electron microscopy. *Scanning Microsc.* **2** 1265–71.

Mohr, D. W., Fritz, S. J. and Eckert, J. O. (1990) Estimation of elemental microvariation within minerals analyzed by the microprobe: use of model population estimates. *Amer. Mineral.* **75** 1406–14.

Morgan, G. B. and London, D. (1996) Optimizing the electron microprobe analysis of hydrous alkali aluminosilicate glass. *Amer. Mineral.* **81** 1176–85.

(2005) Effect of current density on the electron microprobe analysis of alkali aluminosilicate glasses. *Amer. Mineral.* **90** 1131–8.

Moskowitz, B. M., Halgedahl, S. L. and Lawson, C. A. (1988) Magnetic domains on unpolished and polished surfaces of titanium-rich titanomagnetites. *J. Geophys. Res.* **93** 3372–86.

Nash, W. P. (1992) Analysis of oxygen with the electron microprobe: applications to hydrated glasses and minerals. *Amer. Mineral.* **77** 453–7.

Newbury, D. E. (2002) X-ray microanalysis in the variable pressure (environmental) scanning electron microscope. *J. Res. Nat. Inst. Stand. Technol.* **107** 567–603.

Newbury, D. E., Joy, D. C., Echlin, P., Fiori, C. E. and Goldstein, J. I. (1986) *Advanced Scanning Electron Microscopy and X-Ray Microanalysis* (New York: Plenum Press).

Nicholls, J. and Stout, M. Z. (1986) Electron beam analytical instruments and the determination of modes, spatial variations in minerals and textural features of rocks in polished section. *Contrib. Mineral. Petrol.* **94** 395–404.

Nielsen, C. H. and Sigurdsson, H. (1981) Quantitative methods for electron microprobe analysis of sodium in natural and synthetic glasses. *Amer. Mineral.* **66** 547–52.

Oliveira, D. P. S. de, Reed, R. M., Milliken, K. L. *et al.* (2003) (Meta)cherts, (meta)lydites, (meta)phthanites and quartzites of the *série negra* (Crato-S. Martinho), E. Portugal: towards a correct nomenclature based on mineralogy and cathodoluminescence studies. *Ciências da Terra*, special issue no. V, 29.

Pagel, M., Barbin, V., Blanc, P. and Ohnenstetter, D. (2000) *Cathodoluminescence in Geosciences* (Berlin: Springer).

Patsoules, M. G. and Cripps, J. C. (1983) A quantitative analysis of chalk pore geochemistry using resin casts. *Energy Sources* **7** 15–31.

Perkins, W. T. and Pearce, N. J. G. (1995) Mineral microanalysis by laserprobe inductively coupled plasma mass spectrometry. In *Microprobe Techniques in the Earth Sciences*, ed. P. J. Potts, J. F. W. Bowles, S. J. B. Reed and M. R. Cave (London: Chapman and Hall) pp. 291–325.

Pingitore, N. E., Meitzner, G. and Love, K. M. (1997) Discrimination of sulfate from sulfide in carbonates by electron probe microanalysis. *Carbonates Evaporites* **12** 130–3.

Potts, P. J. and Tindle, A. G. (1989) Analytical characteristics of a multilayer dispersion element ($2d = 60$ Å) in the determination of fluorine in minerals by electron microprobe. *Mineral. Mag.* **53** 357–62.

Potts, P. J., Tindle, A. G. and Isaacs, M. C. (1983) On the precision of electron microprobe data: a new test for the homogeneity of mineral standards. *Amer. Mineral.* **68** 1237–42.

Prior, D. J., Boyle, A. P., Brenker, F. *et al.* (1999) The application of electron backscatter diffraction and orientation contrast imaging in the scanning electron microscope to textural problems. *Amer. Mineral.* **84** 1741–59.

Prior, D. J., Trimby, P. W., Weber U. D. and Dingley, D. J. (1996) Orientation contrast imaging of microstructures in rocks using forescatter detectors in the scanning electron microscope. *Mineral. Mag.* **60** 859–69.

Purvis, K. (1991) Fibrous clay mineral collapse produced by beam damage during scanning electron microscopy. *Clay Mineral.* **26** 141–5.

Pyle, J. M., Spear, F. S., Wark, D. A., Daniel, C. G. and Storm, L. C. (2005) Contribution to precision and accuracy of monazite microprobe ages. *Amer. Mineral.* **90** 547–77.

Pyman, M. A. F., Hillyer, J. W. and Posner, A. M. (1978) The conversion of X-ray intensity ratios to compositional ratios in the electron probe analysis of small peaks using mineral standards. *Clays Clay Mineral.* **26** 296–8.

Reay, A., Johnstone, R. A. and Kawachi, Y. (1989) Kaersutite, a possible international microprobe standard. *Geostand. Newslett.* **13** 187–90.

Reed, R. M. and Milliken, K. L. (2003) How to overcome imaging problems associated with carbonate minerals on SEM-based cathodoluminescence systems. *J. Sed. Res.* **73** 328–32.

Reed, S. J. B. (2000) Quantitative trace analysis by wavelength-dispersive EPMA. *Mikrochim. Acta* **132** 145–51.

Reed, S. J. B. and Buckley, A. (1996) Virtual WDS. *Mikrochim. Acta Suppl.* **13**, 479–83.

(1998) Rare-earth element determination in minerals by electron-probe microanalysis: application of spectrum synthesis. *Mineral. Mag.* **62** 1–8.

Rehbach, W. P. and Karduck, P. (1992) *Mikrochim. Acta Suppl.* **121** 153–60.

Reid, A. F., Gottlieb, P., MacDonald, K. J. and Miller P. R. (1985) QEM*SEM image analysis of ore minerals: volume fraction, liberation, and observational variances. In *Applied Mineralogy*, ed. W. C. Park, W. M. Hausen and R. D. Hagni (New York: Metallurgical Society AIME) pp. 191–204.

Reid, A. M., leRoex, A. P., and Minter, W. E. L. (1988) Composition of gold grains in the Vaal Placer, Klerksdorp, South Africa. *Mineral. Deposit.* **23** 211–7.

Richard, L. R. and Clarke, D. B. (1990) AMPHIBOL: a program for calculating structural formulae and for classifying and plotting chemical analyses of amphiboles. *Amer. Mineral.* **75** 421–3.

Robinson, B. W. (1998) The "Geosem" (low-vacuum SEM): an under-utilized tool for mineralogy. In *Modern Approaches to Ore and Environmental Mineralogy*, ed. L. J. Cabri and D. J. Vaughan, Short Course Series, vol. 27 (Ottawa: Mineralogical Association of Canada) pp. 139–51.

Robinson, B. W., Ware, N. G. and Smith, D. G. W. (1998) Modern electron microprobe trace element analysis in mineralogy. In *Modern Approaches to Ore and Environmental Mineralogy*, ed. L. J. Cabri and D. J. Vaughan, Short Course Series, vol. 27 (Ottawa: Mineralogical Association of Canada) pp. 153–80.

Roeder, P. L (1985) Electron-microprobe analysis of minerals for rare-earth elements: use of calculated peak-overlap corrections. *Canad. Mineral.* **23** 263–71.

Schumacher, J. C. (1991) Empirical ferric iron corrections: necessity, assumptions, and effects on selected geothermobarometers. *Mineral. Mag.* **55** 3–18.

Schwartz, A. J., Kumar, M. and Adams, B. L. (2000) *Electron Backscatter Diffraction in Materials Science* (New York: Kluwer).

Sela, J. and Boyde, A. (1977) Cyanide removal of gold from SEM specimens. *J. Microsc.* **111** 229–31.

Small, J. A., Newbury, D. E. and Myklebust, R. L. (1979) Analysis of particles and rough samples by FRAME P, a ZAF method incorporating peak-to-background measurements. In *Microbeam Analysis – 1979*, ed. D. E Newbury (San Francisco, CA: San Francisco Press) pp. 243–6.

Smart, P. and Tovey, N. K. (1982) *Electron Microscopy of Soils and Sediments: Techniques* (Oxford: Oxford University Press).

Smith, D. G. W. and Leibowitz, D. (1986) MinIdent: a database for minerals and a computer program for their identification. *Canad. Mineral.* **24** 695–708.

Smith, J. V. and Rivers, M. L. (1995) Synchrotron X-ray microanalysis. In *Microprobe Techniques in the Earth Sciences*, ed. P. J. Potts, J. F. W. Bowles, S. J. B. Reed and M. R. Cave (London: Chapman and Hall) pp. 163–233.

Smith, M. P. (1986) Silver coating inhibits electron microprobe beam damage of carbonates. *J. Sed. Petrol.* **56** 560–1.

Spear, F. S. and Daniel, C. G. (1998). 3-Dimensional imaging of garnet porphyroblast sizes and chemical zoning: nucleation and growth history in the garnet zone. *Geol. Mater Res.* **1** 1–44.

Statham, P. J. and Pawley, J. (1977) A new method for particle X-ray microanalysis based on peak-to-backround measurement. *Scanning Electron Microsc.* **1978/I** 445–54.

Stormer, J. C., Pierson, M. L. and Tacker, R. C. (1993) Variation of F and Cl X-ray intensity due to anisotropic diffusion in apatite during electron microprobe analysis. *Amer. Mineral.* **78** 641–8.

Tindle, A. G. and Webb, P. C. (1990) Estimation of lithium contents in trioctahedral micas using microprobe data: application to micas from granitic rocks. *Eur. J. Mineral.* **2** 595–610.

(1994) PROBE-AMPH – a spreadsheet program to classify microprobe-derived amphibole analyses. *Computers Geosci.* **20** 1201–28.

Uwins, P. J. R., Baker, J. C. and Mackinnon, I. D. R. (1993) Imaging fluid/solid interactions in hydrocarbon reservoir rocks. *Microsc. Res. Tech.* **25** 465–73.

Waldron, K., Lee, M. R. and Parsons, I. (1994) The microstructures of perthitic alkali feldspars revealed by hydrofluoric acid etching. *Contrib. Mineral. Petrol.* **116** 360–4.

Walker, B. M. (1978) Chalk pore geometry using resin pore casts. In *Scanning Electron Microscopy in the Study of Sediments*, ed. W. B. Whalley (Norwich: Geo Abstracts).

Wallace, P. J. and Carmichael, I. S. E. (1994) S speciation in submarine basaltic glasses as determined by measurements of S Kα X-ray wavelength shifts. *Amer. Mineral.* **79** 161–7.

Ware, N. G. (1991) Combined energy-dispersive–wavelength-dispersive quantitative electron microprobe analysis. *X-Ray Spectrom.* **20** 73–9.

Watt, G. R., Griffin, B. J. and Kinny, P. D. (2000) Charge contrast imaging of geological materials in the environmental scanning electron microscope. *Amer. Mineral.* **85** 1784–94.

Watt, G. R., Oliver, N. H. S. and Griffin, B. J. (2000) Evidence for reaction-induced microfracturing in granulite facies migmatites. *Geology* **28** 327–30.

Watt, G. R., Wright, P., Galloway, S. and McLean, C. (1997) Cathodoluminescence and trace element zoning in quartz phenocrysts and xenocrysts. *Geochim. Cosmochim. Acta* **61** 4337–48.

Wiens, R. C., Burnett, D. S., Armstrong, J. T. and Johnson, M. L. (1994) A simple method to recognize and correct for surface roughness in scanning electron microscope energy-dispersive spectroscopy. *Microbeam Anal.* **3** 117–24.

Willich, P. and Obertop, D. (1990) Quantitative EPMA of ultra-light elements in non-conducting materials. In *Proc. 12th ICXOM, Cracow*, ed. S. Jasleńska and J. Maksymowicz (Kraków: Academy of Mining Metallurgy) pp. 100–3.

Index